Collins

Cambridge Lower Secondary

Science

STAGE 9: WORKBOOK

Heidi Foxford, Beverly Rickwood,
Aidan Gill, Dorothy Warren

Collins

William Collins' dream of knowledge for all began with the publication of his first book in 1819.

A self-educated mill worker, he not only enriched millions of lives, but also founded a flourishing publishing house. Today, staying true to this spirit, Collins books are packed with inspiration, innovation and practical expertise. They place you at the centre of a world of possibility and give you exactly what you need to explore it.

Collins. Freedom to teach.

Published by Collins
An imprint of HarperCollins*Publishers*
The News Building
1 London Bridge Street
London
SE1 9GF

HarperCollinsPublishers
Macken House, 39/40 Mayor Street Upper,
Dublin 1, D01 C9W8, Ireland

Browse the complete Collins catalogue at
www.collins.co.uk

10 9 8 7 6 5

ISBN 978-0-00-836433-5

MIX
Paper | Supporting responsible forestry
FSC™ C007454
www.fsc.org

This book contains FSC™ certified paper and other controlled sources to ensure responsible forest management.

For more information visit: www.harpercollins.co.uk/green

End-of-chapter questions and sample answers have been written by the authors. These may not fully reflect the approach of Cambridge Assessment International Education.

British Library Cataloguing-in-Publication Data

A catalogue record for this publication is available from the British Library.

Authors: Heidi Foxford, Beverly Rickwood, Aidan Gill, Dorothy Warren
Development editors: Anna Clark, Lynette Woodward, Tony Wayte, Sarah Binns
Product manager: Joanna Ramsay
Content editor: Tina Pietron
Project manager: Amanda Harman
Copyeditors: Debbie Oliver, Naomi Mackay
Proofreader: Jan Schubert
Safety checker: Joe Jefferies
Illustrator: Jouve India Private Limited
Cover designer: Gordon MacGilp
Cover artwork: Maria Herbert-Liew
Internal designer: Jouve India Private Limited
Typesetter: Jouve India Private Limited
Production controller: Lyndsey Rogers

Printed in India by Multivista Global Pvt.Ltd.

The publishers gratefully acknowledge the permission granted to reproduce the copyright material in this book. Every effort has been made to trace copyright holders and to obtain their permission for the use of copyright material. The publishers will gladly receive any information enabling them to rectify any error or omission at the first opportunity.

Acknowledgements

The publishers wish to thank the following for permission to reproduce photographs. Every effort has been made to trace copyright holders and to obtain their permission for the use of copyright materials. The publishers will gladly receive any information enabling them to rectify any error or omission at the first opportunity.
(t = top, c = centre, b = bottom, r = right, l = left)

p 15 jinyu36/Shutterstock, p 18 La Gorda/Shutterstock, p 41l FotoRequest/Shutterstock, p 41r Warren Metcalf/Shutterstock, p 49 adapted from Designua/Shutterstock p 62 adapted from VectorMine/Shutterstock, p 64 & p73 Emre Terim/Shutterstock, p 64 Andris Torms/Shutterstock, p 145 Francesco Carucci/Shutterstock, p 150 Andrea Danti/Shutterstock, p 170 Cristian Cestaro/Shutterstock, p 173 Christopher Grimmer/Shutterstock, p 174 NASA images/Shutterstock.

Contents

How to use this book

The questions in this workbook give you an opportunity to continue to learn and test your knowledge and recall at home.

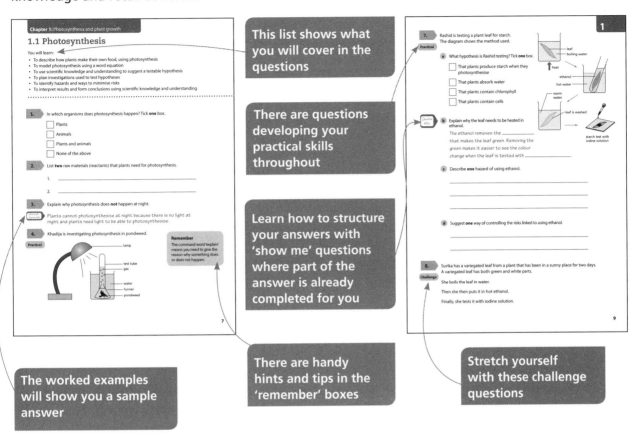

This list shows what you will cover in the questions

There are questions developing your practical skills throughout

Learn how to structure your answers with 'show me' questions where part of the answer is already completed for you

The worked examples will show you a sample answer

There are handy hints and tips in the 'remember' boxes

Stretch yourself with these challenge questions

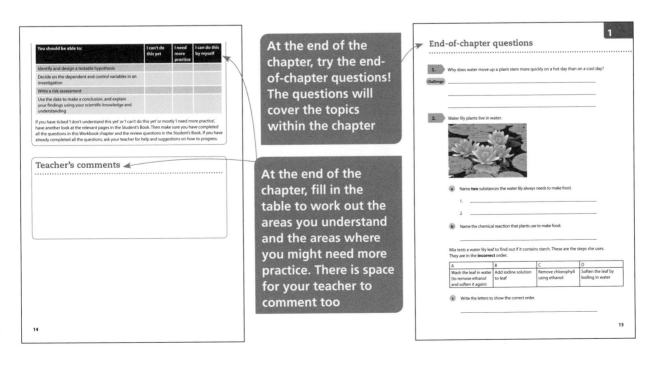

At the end of the chapter, try the end-of-chapter questions! The questions will cover the topics within the chapter

At the end of the chapter, fill in the table to work out the areas you understand and the areas where you might need more practice. There is space for your teacher to comment too

Biology

1.1 Photosynthesis

You will learn:

- To describe how plants make their own food, using photosynthesis
- To model photosynthesis using a word equation
- To use scientific knowledge and understanding to suggest a testable hypothesis
- To plan investigations used to test hypotheses
- To identify hazards and ways to minimise risks
- To interpret results and form conclusions using scientific knowledge and understanding

1. In which organisms does photosynthesis happen? Tick **one** box.

☐ Plants

☐ Animals

☐ Plants and animals

☐ None of the above

2. List **two** raw materials (reactants) that plants need for photosynthesis.

1. _____

2. _____

3. Explain why photosynthesis does **not** happen at night.

 Plants cannot photosynthesise at night because there is no light at night and plants need light to be able to photosynthesise.

4. Khadija is investigating photosynthesis in pondweed.

Practical

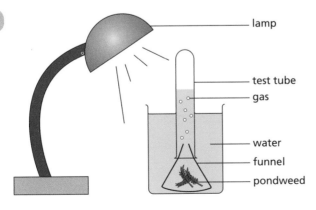

lamp — test tube — gas — water — funnel — pondweed

Remember

The command word 'explain' means you need to give the reason why something does or does not happen.

Bubbles of gas are produced.

a What gas is produced? _____

b How could Khadija test for the gas? _____

c What is the dependent variable in Khadija's investigation?

_____

5. From what is most of a plant's carbohydrate made? Tick **one** box.

☐ Soil

☐ Carbon dioxide and water

☐ Nutrients (from the soil)

☐ Water only

Remember
An important compound that plants make is a large carbohydrate called starch.

6. The diagram below shows a section through a leaf.

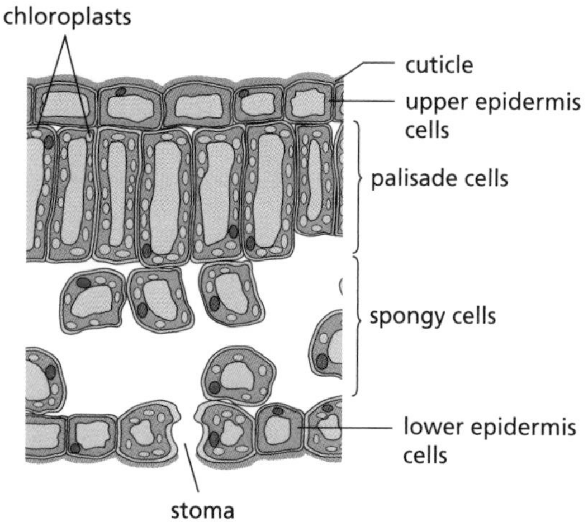

chloroplasts

cuticle

upper epidermis cells

palisade cells

spongy cells

lower epidermis cells

stoma

Use the diagram to help you answer the questions.

a In which part of the leaf does most photosynthesis take place?

_____

b Which part of the leaf has air spaces so that gases can move around?

_____

c Through which part does water escape?

_____

7.

Practical

Rashid is testing a plant leaf for starch.
The diagram shows the method used.

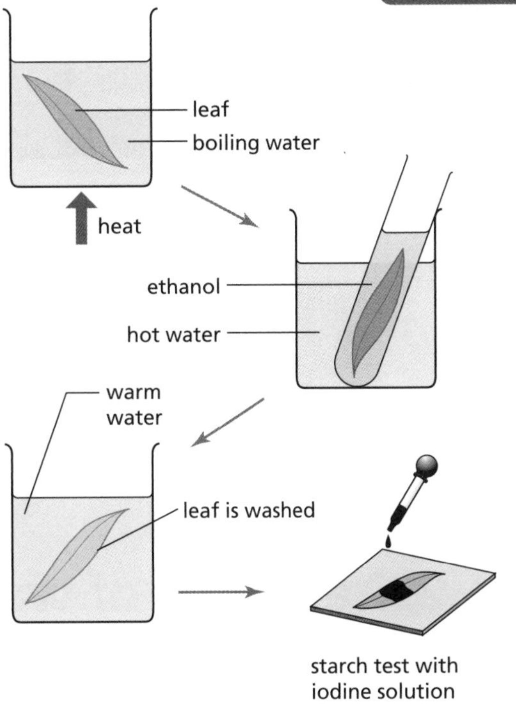

leaf
boiling water
heat
ethanol
hot water
warm water
leaf is washed

starch test with iodine solution

a What hypothesis is Rashid testing? Tick **one** box.

☐ That plants produce starch when they photosynthesise

☐ That plants absorb water

☐ That plants contain chlorophyll

☐ That plants contain cells

Show Me

b Explain why the leaf needs to be heated in ethanol.

The ethanol removes the _____
that makes the leaf green. Removing the
green makes it easier to see the colour
change when the leaf is tested with _____.

c Describe **one** hazard of using ethanol.

_____

_____

_____

d Suggest **one** way of controlling the risks linked to using ethanol.

_____

_____

8.

Challenge

Surika has a variegated leaf from a plant that has been in a sunny place for two days.
A variegated leaf has both green and white parts.

She boils the leaf in water.

Then she puts it in hot ethanol.

Finally, she tests it with iodine solution.

Complete the table to show what results you would expect and why.

	Green part of variegated leaf	White part of variegated leaf
Result of starch test using iodine		
Explanation		

1.2 Transport of water and mineral salts in plants

You will learn:

- To describe the importance of elements found in mineral salts (magnesium and nitrates) for the growth and development of plants
- To use scientific knowledge and understanding to suggest a testable hypothesis

1. Give **one** reason why a plant needs water.

_____

2. Which **two** are functions of root hair cells? Tick **two** boxes.

☐ To absorb glucose

☐ To absorb water

☐ To absorb mineral salts

☐ To absorb carbon dioxide

3. Draw **one** line to match each substance from a mineral salt to its function. Draw **two** lines only.

Substances from mineral salts

Function

Nitrates

To make chlorophyll

To make proteins for growth and repair

Magnesium

To make glucose

4. The diagram shows a root hair cell.

Show Me

Explain **one** way that the root hair cell is adapted to its function.

Its shape gives it a large _____ so it

can absorb _____ more efficiently.

cytoplasm — cell wall — nucleus — vacuole — cell membrane

5. Gabriella forgets to water her houseplant for three weeks. The leaves become floppy. Explain why.

_____

_____

6. The stem of a plant contains tubes made of dead xylem cells. Name **two** substances transported in the xylem.

1._____

2._____

7. Pierre puts the stem of a white flower in a pot of red food colouring mixed with water. After 24 hours the white flower turns pink. Explain why.

_____

_____

8. Xylem cells have thick walls stiffened with lignin. Explain how this feature makes xylem adapted to transporting substances.

_____

_____

9. Some trees have small needles instead of leaves. The needles have fewer stomata and a thick waxy cuticle.

Show Me

Explain **one** way this helps the tree survive.

Fewer stomata and a thick waxy cuticle help the tree lose less _____.

This is important for survival because water is needed for _____.

10. Lana investigates the how air temperature affects the rate of transpiration in a plant. She sets up the equipment shown below.

Challenge

Lana measures how far the bubble moves up the capillary tube in different air temperatures.

a Write the testable hypothesis that Lana is investigating.

_____

_____

b List **one** control variable for Lana's investigation.

_____

rubber tubing

capillary tube

bubble in water

water

Lana's results are shown in the table.

Air temperature (°C)	Distance moved by air bubble in capillary tube in one hour (mm)
15	2
25	4
35	9

Lana only had time to take one measurement at each temperature.

c Suggest what Lana could do to check her measurements were reliable.

_____

_____

_____

_____

Self-assessment

Tick the column which best describes what you know and what you are able to do.

What you should know:	I don't understand this yet	I need more practice	I understand this
Photosynthesis can be modelled using a word equation: carbon dioxide + water → glucose + oxygen			
Inside chloroplasts, chlorophyll traps some energy from light, which is used to power photosynthesis			
Plants use glucose to make the other compounds that they need, including starch (another carbohydrate). Starch is used as a store of energy			
Gases (such as carbon dioxide, oxygen and water vapour) diffuse in and out of open stomata			
Plants need water to make their own food (by photosynthesis) and for cells to keep their shape			
A plant needs mineral salts to make substances			
Plants need magnesium to make chlorophyll and nitrates to make protein			
Water and mineral salts are absorbed by a plant using root hair cells			
Water and mineral salts are transported in tubes formed by dead xylem cells			
Water is lost from a plant through its stomata			
The loss of water vapour through the stomata on the surface of leaves is called transpiration			

You should be able to:	I can't do this yet	I need more practice	I can do this by myself
Identify and design a testable hypothesis			
Decide on the dependent and control variables in an investigation			
Write a risk assessment			
Use the data to make a conclusion, and explain your findings using your scientific knowledge and understanding			

If you have ticked 'I don't understand this yet' or 'I can't do this yet' or mostly 'I need more practice', have another look at the relevant pages in the Student's Book. Then make sure you have completed all the questions in this Workbook chapter and the review questions in the Student's Book. If you have already completed all the questions, ask your teacher for help and suggestions on how to progress.

Teacher's comments

End-of-chapter questions

1.

Why does water move up a plant stem more quickly on a hot day than on a cool day?

Challenge

_____

_____

_____

2.

Water lily plants live in water.

a Name **two** substances the water lily always needs to make food.

1. _____

2. _____

b Name the chemical reaction that plants use to make food.

_____

Mia tests a water lily leaf to find out if it contains starch. These are the steps she uses. They are in the **incorrect** order.

A	B	C	D
Wash the leaf in water (to remove ethanol and soften it again)	Add iodine solution to leaf	Remove chlorophyll using ethanol	Soften the leaf by boiling in water

c Write the letters to show the correct order.

_____

d Suggest why you find few plants growing underneath water lillies.

_____

_____

_____

e Explain how a water lily is different to a human in the way it gets its nutrition.

_____

_____

_____

3.

Practical

Carlos has a celery stalk with leaves. He cuts the bottom off the stalk. Then he puts the stalk in a glass of water mixed with blue food colouring.

Next morning, the celery leaves are blue. Carlos cuts the stalk halfway up and sees small blue circles within the stalk.

a Explain his observations.

_____

_____

_____

b What are the blue circles inside the stalk?

_____

_____

c Why have the leaves turned blue?

_____

_____

d Carlos's friend says that if you cut the leaves off a celery stalk, less water will be taken in by the stalk.

Plan an investigation method to find out if this might be true.

_____

_____

_____

_____

_____

4. Hussain is growing plants. He waters them every day. Once a week he adds fertiliser to the soil.

Name the specialised cells in the plants that:

a absorb water and minerals _____

b transport water and minerals up the stem to the leaves.

_____

c If plants do not get enough nitrates they don't grow properly. Explain why.

_____

_____

d Hussain removes the roots from a plant. Explain why the plant could die after the roots are removed.

_____

_____

2.1 Producing urine

You will learn:

- To describe the structure of the human excretory (renal) system
- To describe how the kidneys filter blood to remove urea, which is excreted in urine
- To describe the uses, strengths and limitations of models and analogies
- To evaluate experimental methods, explaining any improvements suggested

1. The diagram below shows the organs of the human excretory system.

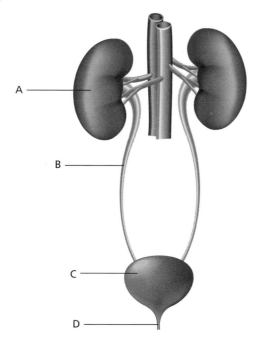

a Draw **one** line to match each organ name to the correct letter label.

Organ	Correct letter label
Bladder	A
Kidney	B
Ureter	C
Urethra	D

b Write the letter of the organ that stores urine.

_____

c What is the function of **D**?

_____

2. Which of the following statements best describes excretion?
Tick **one** answer only.

The removal of solid waste from an organism ☐

Getting rid of waste substances from an organism ☐

The removal of water from an organism ☐

Breathing out carbon dioxide ☐

3. Describe the function of the kidneys.

Show Me

The kidneys filter the _____ and remove wastes such as

_____ .

4. Name a substance that is excreted by the lungs.

_____

5. The excretion of urine involves several steps. Complete the table using numbers 1–4 to put the steps in the correct order.

The first one has been done for you.

The kidney filters urea and water from the blood to form urine	1
Urine travels to the outside of the body in the urethra	
Urine travels in the ureter	
Urine is stored in the bladder	

6. Plants excrete oxygen during photosynthesis.

Challenge Explain why the release of oxygen gas from a plant is considered to be excretion.

_____

_____

7. Devraj has kidney disease. His blood contains too much urea because his kidneys are not functioning properly.

Give the reason why too much urea in the blood could be dangerous for Devraj.

_____

8.

Practical

Scientists can find out how well a person's kidneys are working by testing a urine sample for the presence of protein.

A scientist is testing some urine samples from patients. The scientist tests each urine sample for protein using Biuret solution.

A label on the Biuret solution shows this hazard symbol.

a State **one** way the scientist could control the risks when using Biuret solution.

_____

The table shows the scientist's results

Urine sample	Colour change with Biuret solution
A	No change
B	Turned purple
C	Turned purple
D	No change

b Which **two** samples contained protein?

_____

c Stress can cause small amounts of protein in the urine.

People with kidney disease often have high levels of protein in their urine.

Suggest a further investigation to find out if the two urine samples containing protein were from patients suffering from stress or patients with kidney disease.

_____

_____

9.

Challenge

Anand writes an analogy for the function of the kidneys:

'The kidneys are like a swimming pool filter that removes dirt and debris from the water.'

Suggest **one** strength and **one** limitation of Anand's analogy.

_____

_____

Self-assessment

Tick the column which best describes what you know and what you are able to do.

What you should know:	I don't understand this yet	I need more practice	I understand this
The human excretory system contains the kidneys, ureters, bladder and urethra			
The kidneys filter the blood and remove wastes, such as urea			
The wastes are dissolved in water and form urine, which passes down the ureters to the bladder			
The bladder stores urine until it can be emptied			

You should be able to:	I can't do this yet	I need more practice	I can do this by myself
Suggest strengths and weaknesses for analogy models			
Write risk assessments			
Suggest further investigations that can be done based on conclusions from an investigation			

If you have ticked 'I don't understand this yet' or 'I can't do this yet' or mostly 'I need more practice', have another look at the relevant pages in the Student's Book. Then make sure you have completed all the questions in this Workbook chapter and the review questions in the Student's Book. If you have already completed all the questions, ask your teacher for help and suggestions on how to progress.

Teacher's comments

End-of-chapter questions

1. Complete the sentences using words from the box.

brain	cells	liver	useful	waste

The excretory system gets rid of _____ substances the body has

made inside its _____ .

2. Name **two** organs found in the excretory system.

_____

3. What is another name for the excretory system?

_____

4. Explain the function of each of these organs:

a bladder: _____

b kidney: _____

5. How does urine flow? Tick **one** box only.

From the bladder to the kidney ☐

From the kidney to the bladder ☐

From the stomach to the kidney ☐

From the ureter to the liver ☐

6. List **two** substances that are excreted by humans.

_____

_____

7. Name the part of the blood that carries the waste products to the kidneys.

_____

3.1 Variation within a species

You will learn:

• To describe how genetic differences between organisms can lead to variation within the species

....

1.   Complete the sentences using words from the list.

discontinuous   **continuous**   **group**   **kingdom**

A characteristic that changes gradually over a range of values is said to have _____
variation. A characteristic that has a distinct range of options or categories is said to have
_____ variation.

2.   Draw a line to match each characteristic to the correct type of data.

Characteristic   **Type of data**

Height

Arm span   Discontinuous

Able to roll tongue   Continuous

Foot size

3.   The table below gives descriptions of features of a hydrangea plant. Tick the
correct column to show whether each feature is an example of inherited or
environmental variation.

Description of feature	Example of inherited variation	Example of environmental variation
Flat, pointed leaves		
Leaves that have turned yellow due to lack of magnesium in soil		
Blue flowers due to acidic soil		

4. Scientists can classify human blood as being one of four different types, called 'blood groups'.

A scientist investigates the number of people with different blood groups. Her results are shown in the chart below.

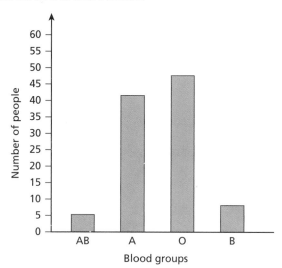

a How many people had the blood group AB? _____

b Which was the most common blood group? _____

Show Me

c Explain why the graph has been drawn with spaces between the bars.

Because blood groups are a type of _____ variation and this

type of data should always be presented with _____ .

5. Explain why human skin colour is an example of both inherited variation and environmental variation.

_____

_____

6. Look at the data below, which shows the individual weights of a group of students.

41.7 kg   43.3 kg   39.2 kg   34.8 kg   43.3 kg   40.8 kg,   46.7 kg   41.5 kg   42.2 kg   35.9 kg

a State whether the data is an example of continuous or discontinuous variation.

_____

> **Remember**
> The range is the lowest and highest values in a set of data.

b State the range of the data. _____

7.

Practical

Lily is measuring the length of leaves from a tree.

She measures the length of 60 leaves and records her results in a tally chart.

Her results are shown below:

Grouped lengths of leaves (mm)	Tally	Total
41–50	////	
51–60	₩₩ ₩₩	
61–70	₩₩ ₩₩ ₩₩ /	
71–80	₩₩ ₩₩ ₩₩ ₩₩	
81–90	₩₩ ////	
91–100	/	

a Complete the tally chart by adding the total for each group of measurements.

b Draw a bar chart to show the data.

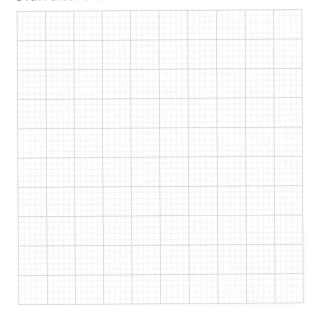

Remember
A bar chart showing continuous data should **not** have gaps between each bar.

c Lily finds a leaf which is 150 mm long. Explain why it is unlikely that this leaf comes from the tree that she investigated.

_____

Challenge **d** Explain why it is impossible to know the true **range** of Lily's set of data.

_____

_____

3.2 Chromosomes, genes and DNA

You will learn:

- To describe how a chromosome contains a molecule of DNA, some sections of which are genes
- To describe the role of genes in inherited variation
- To describe the uses, strengths and limitations of models and analogies

1. Which statement **best** describes DNA? Tick **one** box only.

☐ A molecule that carries genetic information

☐ An acid released by the brain

☐ A substance that controls growth

☐ A molecule that carries oxygen

2. In which part of a cell is DNA found?

_____

3. Draw lines to match each term to its correct definition.

Genetic material   A section of DNA that controls the development of a particular characteristic

Gene   A structure made from a molecule of DNA folded around proteins

Chromosome   A substance found in a cell that controls how the cell develops

4. The diagram shows the genetic material found inside the nucleus of a cell.

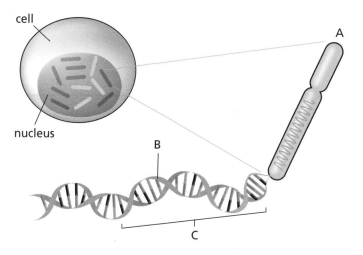

Remember
It is easy to get confused with DNA, genes and chromosomes so make sure you know the difference.

a Which letter shows a chromosome?

b Which letter shows DNA?

_____

c **C** shows a section of genetic material that controls the development of a specific characteristic. What is the name of **C**?

_____

> Worked
> Example

d All rabbits have 44 chromosomes. However, rabbits have many different fur colours.

Explain how a rabbit's genes result in inherited variation in their fur colour.

The genes on the rabbit's chromosomes have slight differences between them in each individual. For example, one rabbit might inherit the genes for brown fur while another rabbit might inherit the gene for black fur. This results in the variation in fur colour.

5. In the 1940s, scientists knew that DNA was a very important molecule. However, they did not know about its structure.

In the 1950s, two scientists, Franklin and Wilkins, studied DNA using X-rays. Franklin and Wilkins were experts in a technique called X-ray diffraction.

Franklin produced an X-ray photograph that gave important clues about the structure of DNA. This allowed two other scientists, Watson and Crick, to produce a 3D model of DNA.

a What question were all the scientists trying to answer?

_____

b Give **one** piece of evidence that Watson and Crick used to produce their model.

_____

c Explain how the work of Watson and Crick was made possible by other scientists.

_____

_____

Challenge **d** Watson and Crick produced a 3D model of DNA. Explain why scientists often use models to represent their ideas in science.

_____

_____

3.3 Fertilisation and inheritance

You will learn:

- To describe how fertilisation in humans occurs
- To explain why organisms inherit features from both their parents
- To describe the role of the XX and XY chromosomes in human sex determination

1. What is the purpose of sexual reproduction. Tick **one** box.

☐ To produce energy ☐ To produce hormones

☐ To produce offspring ☐ To help a person grow

2. Complete the table to show the names of the male and female gametes and the number of chromosomes they contain in their nucleus.

	Name of gamete	Number of chromosomes
Males		
Females		

3. What is the name of the process when a sperm cell and egg cell fuse? _____

> **Remember**
> It is important to learn the meaning of all the scientific words in each topic. You should use scientific words in your answers.

4. Describe what happens after the egg cell and sperm cell fuse.

Show Me

The fusion of the egg cell and sperm cell makes a

_____. This forms a ball of cells which

then develops into an _____ .

5. Explain why a sperm cell and egg cell only have half the number of chromosomes found in a normal body cell.

_____

_____

_____

6. Explain why all your body cells contain 46 chromosomes.

_____

_____

7. State the combinations of sex chromosomes that determine the sex of a male and female.

_____

_____

8. Female egg cells contain one X chromosome. Explain how a man's sperm cell determines the sex of the baby.

_____

_____

9. A man and woman have two daughters. Although the daughters are sisters, they do not look identical. Explain why.

Challenge

_____

_____

_____

3.4 Fetal growth and development

You will learn:

- To describe the development of the fetus
- To explain how the fetus is protected and nourished
- To discuss the ways in which a mother's health and lifestyle choices (diet, smoking and drugs) can affect the development of a fetus

1. Complete the sentences using the words from the list.

cell division	cell respiration	decreases	increases	menstruation

When an egg is fertilised, _____ happens to make more cells.

This _____ the size of the embryo, making it grow.

2. What is an embryo? Tick **one** box.

☐ A tiny ball of cells that grows from a fertilised egg

☐ A fetus with a beating heart

☐ An egg cell that is waiting to be fertilised

☐ A specialised cell

3. Give the meanings of the terms 'fertilisation' and 'specialisation'.

_____

_____

4. The table shows the length of a fetus at different times during a pregnancy.

Time in pregnancy (weeks)	9	12	16	20	24	28	32	34	40
Length of fetus (mm)	60	100	140	190	230	270	300	340	380

a Draw a line graph to show how the length of the fetus changes during the pregnancy.

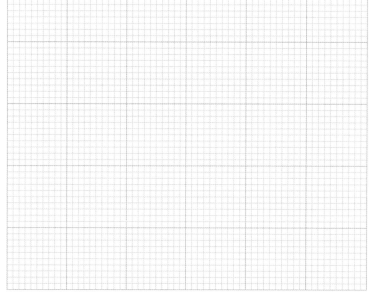

Show Me

b When is the fastest period of growth of the fetus? Explain your answer.

Between _____ because the

line of the graph is _____ during this period.

c When is the slowest period of growth of the fetus?

Explain your answer.

_____

_____

Remember

A steeper slope on a graph shows a greater rate of change.

5.

Challenge

Explain why it is important for a pregnant woman to have a healthy, balanced diet.

_____

_____

6.

Describe the problem caused by a lack of folic acid in the diet of a pregnant woman.

_____

_____

7.

Suggest why doctors may need to be careful about giving medicines to pregnant women.

_____

8.

Describe **two** harmful effects that a pregnant woman can cause to the fetus if she smokes during pregnancy.

1. _____

2. _____

9.

Name **one** of the drugs in tobacco smoke that reduces the amount of oxygen the fetus gets.

_____

_____

10.

Challenge

Explain why a doctor might be concerned if a pregnant woman gets a virus.

_____

_____

Remember

If there are **two** marks for a written question, you will need to make **two** points in your answer.

3.5 Natural selection

You will learn:

- To describe how organisms are adapted to where they live
- To state what is meant by 'evolution'
- To describe how genetic differences between organisms can lead to variation within the species
- To describe how genetic changes over time result in natural selection

1. Draw lines to match the words with their correct definitions.

Word	Definition
Adaptation	the differences that exist between individuals
Variation	characteristic of an organism that allows it to survive in a certain ecosystem
Evolution	a gradual change in the characteristics of an organism over time

2. Arctic hares are adapted to live in the Arctic ecosystem. Complete the table to describe how each adaptation helps the arctic hare survive in the Arctic.

Adaptation	How the feature helps the Arctic hare survive
Small ears	
Strong claws	
White fur	

3. Complete the sentences by choosing words from this list.

> **variation**   **evolution**   **fossil**   **rocks**   **mutation**

Darwin's theory of _____ explains how species have changed over

time. The theory is supported by evidence from observations of organisms and from

_____ records.

4. Charles Darwin caught different types of finches on the Galápagos Islands in the Pacific Ocean. He brought them back to London where they were analysed by John Gould. Each species had a different type of beak. The diagram shows two of these finches.

A   B

strong beak   long slender beak

Complete the table to show:

- the letter of the finch best suited to each type of food
- an explanation of how the beak is adapted to the food.

Finch	Main food source	Explanation
	Nectar from inside flowers	
	Seeds that need to be crushed	

5. Explain what is meant by 'natural selection'.

Organisms that are better _____

have a better chance of surviving and are therefore

more likely to _____ .

Remember

Don't get *natural selection* and *evolution* mixed up. Natural selection is when a characteristic becomes more common in a population because organisms with the characteristic are more likely to survive. Evolution is the gradual changes of a species over time (which happens because of natural selection).

6.

Challenge

Lemmings are small animals naturally found in the Arctic. In 1800 a population of lemmings was taken from the Arctic to an area with a warmer climate. The lemmings were left to live and reproduce in the new area. The population of lemmings was revisited 200 years later.

Some information about the lemmings is shown in the table.

Year	Average fur length of lemming (mm)
1800	12
2000	8

Explain how Darwin's theory of natural selection can explain this change.

_____

_____

_____

_____

Self-assessment

Tick the column which best describes what you know and what you are able to do.

What you should know:	I don't understand this yet	I need more practice	I understand this
Data that shows continuous variation may have any value (within a range)			
Data that shows discontinuous variation has a limited number of options			
Variation that an organism gets from its parents is inherited variation			
Variation that is caused by environmental factors is environmental variation			
DNA is the genetic material that carries information determining how individual plants and animals develop			
Chromosomes are found in the nucleus of the cell and contain DNA			
Genes are sections of DNA that control the development of a specific characteristic and so control an organism's inherited variation			
Male gametes are sperm cells and female gametes are egg cells			
Gametes carry only one copy of each type of chromosome			
When a sperm cell nucleus and an egg cell nucleus fuse, a fertilised cell is made. This develops into an embryo, which develops into a baby			
A fertlised cell contains two copies of each type of chromosome (one from the mother and one from the father)			
In humans, males have XY sex chromosomes and females have XX chromosomes			
A fertilised cell develops into an embryo, which grows into a fetus			
Substances diffuse between the mother's blood and the fetus' blood			

	I can't do this yet	I need more practice	I can do this by myself
The fetus' development can be slowed or harmed by poor diet (e.g. lack of folic acid), drugs (including medicines), smoking and diseases			
Organisms have adaptations that help them to survive in their ecosystems			
Evolution is gradual change over time			
Charles Darwin proposed the theory of evolution by natural selection			
Natural selection happens when there is variation in a species, with some organisms being (by chance) better adapted than others			
Natural selection means that those who are better adapted to their environment tend to be more likely to survive and have more offspring			
Slight differences and natural changes to genes are what cause inherited variation			
Genes that help an organism to survive better than others of the same species are more likely to be passed down to the next generation			

You should be able to:	I can't do this yet	I need more practice	I can do this by myself
Identify the range of a set of data			
Describe the development of scientific understanding through collaboration			
Design and use a tally chart to help count items			
Select and draw the correct type of bar chart to show your continuous and discontinuous data			
Explain the use of models in science			
Compare rates of change on a line graph			

If you have ticked 'I don't understand this yet' or 'I can't do this yet' or mostly 'I need more practice', have another look at the relevant pages in the Student's Book. Then make sure you have completed all the questions in this Workbook chapter and the review questions in the Student's Book. If you have already completed all the questions, ask your teacher for help and suggestions on how to progress.

Teacher's comments

End-of-chapter questions

1. The graph below shows the heights of a population of people.

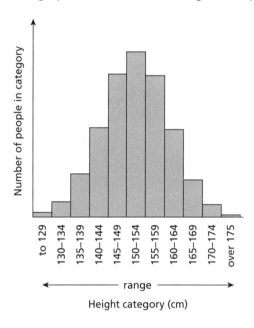

Height category (cm)

a What is the most common height range?

_____

b What type of variation is shown in the graph? Explain your answer.

_____

_____

c   Explain how genes can cause inherited variation in height.

_____

_____

2.   Which of the following **best** describes a gene? Tick **one** box only.

☐ A cell that carries genetic information

☐ A copy of all the DNA found in the body

☐ A damaged section of DNA

☐ A section of DNA that has instructions for a characteristic

3.   Opi and Ken are expecting a baby.

Sperm cells and egg cells are needed for sexual reproduction.

a   What is the general name for a sex cell?

_____

b   Describe the process by which the gametes produce an embryo.

_____

_____

c   Opi reads a leaflet that says that pregnant women should eat a healthy diet.
Explain why.

_____

_____

d   Opi and Ken have a baby girl. Which **one** of the following are the sex
chromosomes found in a female?

XX   XY   XZ   XF

4. Thousands of years ago, the ancestors of giraffes lived in Africa. Some had slightly longer necks than others. They fed off leaves in the trees in their habitat. The giraffes with slightly longer necks were better adapted to survive.

a Suggest **one** reason why giraffes with longer necks survived.

_____

_____

The diagram shows how the giraffes in Africa today have longer necks that their ancestors.

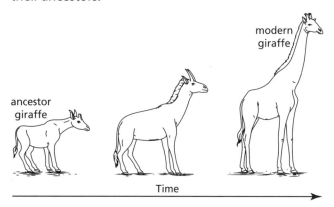

Scientists believe that natural selection could explain the change in the length of the giraffe neck over time.

b Explain how natural selection could have caused the change in the length of the giraffe neck over a long period of time.

_____

_____

_____

4.1 Populations and extinction

You will learn:

- To explain ways in which organisms are adapted to their ecosystems
- To describe the effects of an environmental change on the population of a species (including extinction)

1. Complete the table to show what resources organisms need to survive. Add **one** tick to each row. The first row has been done for you.

Resource	Plants	Animals	Both plants and animals
Sunlight	✓		
Water			
A source of food			
Carbon dioxide			

2. Complete the sentence by choosing words from this list.

temperature ecosystem reproduce move survive respire

An adaptation is a feature of a species that helps it to _____ and

_____ in a particular _____ .

3. Draw lines to match each physical adaptation of a desert tortoise to its correct explanation of how the feature helps it to survive in its ecosystem.

Adaptation

Shell

Wide feet

Strong and muscular legs

How the feature helps

good for walking on sand

provides protection from predators

allows it to burrow into ground to keep cool

4. The barrel cactus grows in the desert.

This plant is adapted to life in the desert. Explain how each adaptation helps the barrel cactus to survive.

Show Me

a Large swollen stem:

stores water so the cactus _____ .

b Thick spines rather than leaves:

_____

_____ .

5. Complete the sentences by choosing words from this list.

temperature	competition	predation	water availability

The size of a population can be changed by living factors or non-living factors. Non-living

factors include _____ , sunlight, _____ and pollution.

Living factors include disease, _____ and _____ .

6. Name **one** physical factor that could limit a plant population at the bottom of a lake.

_____

7.

Challenge

Scientists discovered a sudden decrease in the population of a species of toad in the forests of Costa Rica. The population decreased after the local climate changed: the rainfall decreased and the temperature increased.

Suggest how a change in the local climate could have caused a sudden decrease in the population of toads.

Remember
When a question asks you to 'suggest', think about what you already know and come up with your own ideas about the information you are given.

_____

_____

_____

8. Lynx are predators of snowshoe hares.

snowshoe hare

lynx

The graph shows how the estimated population of snowshoe hares and lynx in one part of Canada changed over 20 years.

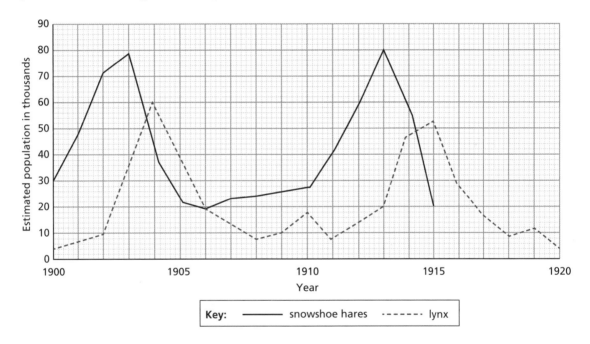

Year	1916	1917	1918	1919	1920
Estimated number of snowshoe hares (in thousands)	28	16	9	12	5

a Use the data shown in the table to finish plotting the line for the snowshoe hares.

b Describe how the population of lynx changed between 1913 and 1917.

_____

c When the snowshoe hare population increased, the lynx population also started to increase. Explain why.

_____

_____

9. What is meant by the term 'extinction'?

_____

10. List **two** causes of extinction of a population.

1. _____

2. _____

11. Some scientists monitored the population of *Cabomba* plants in a lake over a period of 6 months.

Practical

	January	February	March	April	May	June
Number of plants	765	768	772	483	212	167

a Between which months did the number of *Cabomba* begin to decrease?

_____

_____

Challenge **b** The decrease in the population of the *Cabomba* was thought to be caused by an invasive plant species introduced into the lake in the year the data was collected.

Explain how an invasive species could have caused the population of *Cabomba* to decrease.

Remember
An invasive species is a non-native species that may damage an ecosystem.

_____

_____

c List **two** other possible reasons for this decrease.

1. _____

2. _____

Self-assessment

Tick the column which best describes what you know and what you are able to do.

What you should know:	I don't understand this yet	I need more practice	I understand this
Population sizes are altered by changes in physical and living factors in an environment			
Predator and prey populations change constantly when one population depends on the other			
There are many reasons why a species becomes becomes extinct, including permanent changes in physical factors, diseases, competition for resources and catastrophic events			
The adaptations of an organism may mean that it is unable to survive in an area if the conditions change			

You should be able to:	I can't do this yet	I need more practice	I can do this by myself
Find and present ordered information in a suitable graph			

If you have ticked 'I don't understand this yet' or 'I can't do this yet' or mostly 'I need more practice', have another look at the relevant pages in the Student's Book. Then make sure you have completed all the questions in this Workbook chapter and the review questions in the Student's Book. If you have already completed all the questions ask your teacher for help and suggestions on how to progress.

Teacher's comments

End-of-chapter questions

1. The camel is a mammal that is adapted to living in sandy deserts. Two physical adaptations of the camel are:

- large flat feet
- two rows of long eyelashes.

a Explain how each of these adaptations helps the camel to survive.

- Large flat feet: _____

- Two rows of long eyelashes: _____

b Which **one** of the following is a **living** factor that could limit the size of a population of camels in the desert? Tick **one** box.

☐ light ☐ competition ☐ water ☐ temperature

c List **two** physical factors that could affect the size of a population of plants in a desert ecosystem.

_____

_____

d The cactus plant is usually found in hot deserts. It has spikes instead of leaves.

Complete the table to explain how each adaptation helps the cactus to survive in the desert. The first explanation has been done for you.

thick waxy skin

large fleshy stems

spikes

shallow, widespread roots

Feature	How the feature helps the cactus survive in the desert
Thick waxy covering	Slows down water lost from evaporation
Large fleshy stems	
Spikes	
Shallow roots that spread out a long way	

2. The graph shows a predator–prey cycle for wolves and hares. Tick **two** boxes.

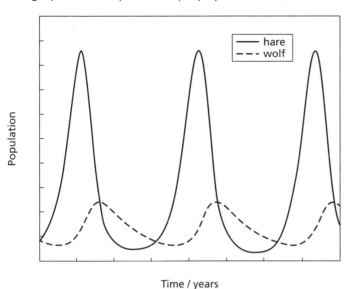

Time / years

a) Which organism is a predator? _____

b) The graph shows that shortly after the population of hares increases, the population of wolves increases. Explain why.

_____

_____

c) Predict what would happen to the wolf population if all the hares got a disease and became extinct.

_____

d) Explain your answer to part **c**.

_____

3. The *Camelops* is a species of extinct camel that lived in North America many years ago. Scientists are investigating factors that could have caused the *Camelops* to become extinct.

a) Which **two** of the following are possible causes of extinction of a species?

☐ hunting   ☐ pollution

☐ habitat protection   ☐ decrease in competition

b) Some scientists think that a change in climate in North America could have caused the *Camelops* to become extinct.

Explain how a permanent change in climate could have caused extinction of the *Camelops*.

_____

_____

Chemistry

5.1 Atomic structure and the Periodic Table

You will learn:

- To understand how the position of an element in the Periodic Table can be used to predict its atomic structure and chemical and physical properties
- To understand that models and analogies can change according to scientific evidence
- To describe the uses, strengths and limitations of models and analogies
- To represent scientific ideas using recognised symbols or formulae

1. Draw a line to match the word with its correct meaning.

The first line has been drawn for you.

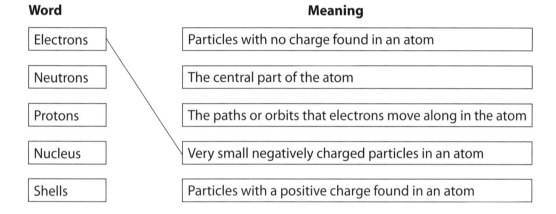

Word		Meaning
Electrons		Particles with no charge found in an atom
Neutrons		The central part of the atom
Protons		The paths or orbits that electrons move along in the atom
Nucleus		Very small negatively charged particles in an atom
Shells		Particles with a positive charge found in an atom

2. Complete the sentences.

Use the words in the box.

atomic	mass	protons	electrons	nucleus
	shells	neutrons	atom	

a An _____ number is the protons an _____ of an element has.

b A _____ number is the total number of _____ and

_____ found in the nucleus of an atom.

3. The diagram shows the model of the atom we used today.

a Describe today's model of the atom.

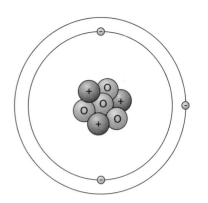

_____

_____

_____

b Explain why scientists use models to describe atoms.

_____

_____

_____

c The model of the atom has changed over time.

Suggest a reason why.

_____

_____

4. The symbol for sulfur in the Periodic Table is $^{32}_{16}$S.

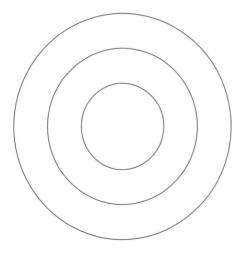

Complete the diagram of the sulfur atom by adding the protons, neutrons and electrons.

5. The diagram shows the model of a magnesium atom.

Worked Example

Use the diagram to explain where magnesium is found in the Periodic Table.

There are three electron shells, so the atom is found in Period 3.

There are two electrons in the outer shell, so the atom is found in Group 2.

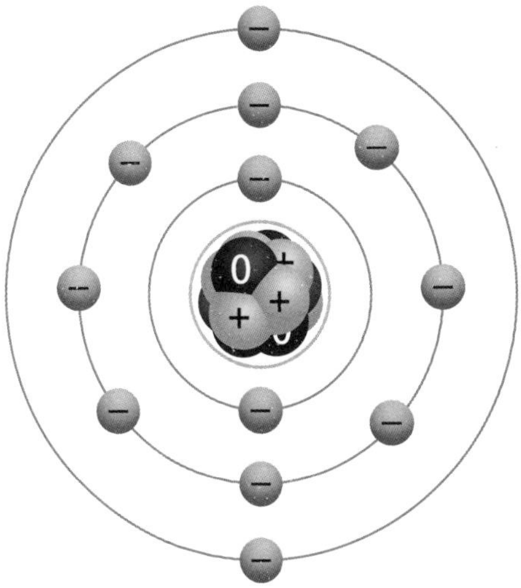

6. The diagram shows the model of a carbon atom.

Use the diagram to explain where carbon is found in the Periodic Table.

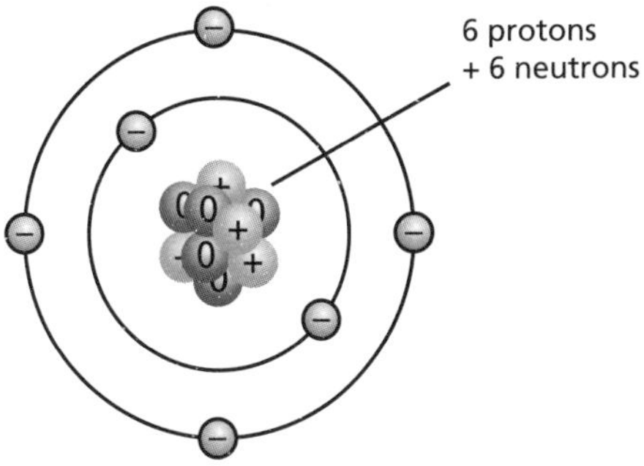

6 protons
+ 6 neutrons

_____

_____

_____

carbon atom

7. **a** Draw a diagram to show the structure of the potassium atom, $^{39}_{19}K$.

Challenge

b Explain how the structure of the atom relates to its position in the Periodic Table.

_____

_____

_____

c Evaluate the model of the atom drawn in **a** by writing down a strength and limitation.

_____

_____

_____

8. The diagram shows an image of Rutherford's model of the atom.

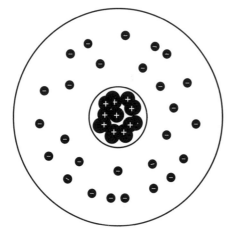

a Compare Rutherford's model of the atom with the one we use today.

_____

_____

_____

b Explain why Rutherford's model of the atom is no longer used by scientists today.

_____

_____

_____

_____

5.2 Trends in the Periodic Table

You will learn:

- To understand that trends in chemical and physical properties are seen within the different groups of the Periodic Table and describe these trends using the Group 1 elements as examples
- To represent scientific ideas using recognised symbols or formulae
- To interpret results, form conclusions using scientific knowledge and understanding and explain the limitations of those conclusions
- To present results appropriately and use them to predict values between the data collected

1. Neon is in Group 8 of the Periodic Table. Argon is below neon in the group.

Which of these statements is **true**? Tick **one** statement.

☐ Neon is more reactive than argon.

☐ Neon is less reactive than argon.

☐ Neon is an inert gas.

☐ Neon is a solid.

2. The diagram shows the outline of the Periodic Table.

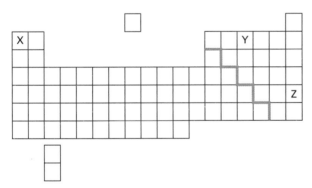

a Shade a group of non-metals blue.

b Shade the Group 2 metals green.

c Draw a red box round the Period 2 elements.

d Which of the three elements indicated **X**, **Y** and **Z** has atoms with one electron in their outer shells?

_____

e Which of the three elements has atoms with the most electron shells?

_____

3. Complete the sentences using words from this list.

| faster | slower | carbon dioxide | oxygen | hydrogen | above | below | next to |

When lithium reacts with water, _____ gas is formed. Sodium is

_____ lithium in their group. The rate of reaction of sodium with

water is _____ than the rate of reaction of lithium and water.

4. Explain the difference between chemical properties and physical properties.

_____

_____

_____

5. Give **two** examples of chemical properties and **two** examples of physical properties.

_____

_____

6. Explain why elements in the same group of the Periodic Table have similar chemical properties.

Challenge

_____

_____

7. This is Mendeleev's Periodic Table from 1869.

I						
H 1.01	II	III	IV	V	VI	VII
Li 6.94	**Be** 9.01	**B** 10.8	**C** 12.0	**N** 14.0	**O** 16.0	**F** 19.0

VIII

Na 23.0	**Mg** 24.3	**Al** 27.0	**Si** 28.1	**P** 31.0	**S** 32.1	**Cl** 35.5			
K 39.1	**Ca** 40.1		**Ti** 47.9	**V** 50.9	**Cr** 52.0	**Mn** 54.9	**Fe** 55.9	**Co** 58.9	**Ni** 58.7
Cu 63.5	**Zn** 65.4			**As** 74.9	**Se** 79.0	**Br** 79.9			
Rb 85.5	**Sr** 87.6	**Y** 88.9	**Zr** 91.2	**Nb** 92.9	**Mo** 95.9		**Ru** 101	**Rh** 103	**Pd** 106
Ag 108	**Cd** 112	**In** 115	**Sn** 119	**Sb** 122	**Te** 128	**I** 127			
Ce 133	**Ba** 137	**La** 139		**Ta** 181	**W** 184		**Os** 194	**Ir** 192	**Pt** 195
Au 197	**Hg** 201	**Ti** 204	**Pb** 207	**Bi** 209					
			Th 232		**U** 238				

Explain why Mendeleev left gaps in his version of the Periodic Table.

_____

_____

8. Jamila and Yuri are investigating the Periodic Table.

Practical They find some data on the internet about a group of non-metals which looks at the relationship between their atomic number and their densities.

Density is a measure of the amount of **mass** something has in a certain **volume**.

The data is shown in the table below.

Element						
Atomic number	2	10	18	36	54	86
Density in g/dm³	0.18	0.90	1.78	3.71	5.85	9.97

a Complete the first row of the table by writing in the names of the elements. To do this you will need to look at a copy of the Periodic Table.

b Jamila thinks that there is a pattern in the data. Plot the data on the graph axes below to find out if she is right. Add a line of best fit.

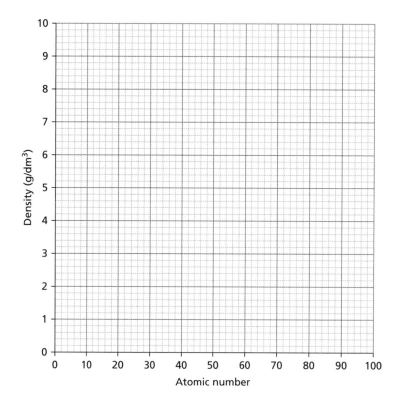

c Describe the pattern between atomic number and density using the information in your graph.

_____

_____

d Use the graph to predict the density of the elements with atomic number:

i 31 (gallium) _____

ii 49 (indium) _____

e Yuri thinks that the predictions for the density of gallium and indium are wrong.

Jamila disagrees because she thinks the graph is accurate.

Explain why Yuri is correct.

_____

_____

Self-assessment

Tick the column which best describes what you know and what you are able to do.

What you should know:	I don't understand this yet	I need more practice	I understand this
The model of the atom we currently use has a nucleus with electrons moving round it in shells			
Elements are arranged in order of atomic number in the Periodic Table			
Elements in the same group share similar properties but there are trends, such as reactivity and boiling point, within the groups			

You should be able to:	I can't do this yet	I need more practice	I can do this by myself
Discuss how models can change according to scientific evidence			
Describe the uses, strengths and limitations of models			
Represent elements using symbols and formulae			
Use graphs to predict values between data points			
Interpret results and form conclusions			

If you have ticked 'I don't understand this yet' or 'I can't do this yet' or mostly 'I need more practice', have another look at the relevant pages in the Student's Book. Then make sure you have completed all the questions in this workbook chapter and the review questions have another look at the relevant pages in the Student's Book. If you have already completed all the questions ask your teacher for help and suggestions on how to progress.

Teacher's comments

End-of-chapter questions

1. Look at the diagram of a nitrogen atom.

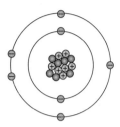

a How many electrons are in the atom? _____

b How many neutrons are in the atom? _____

c How many protons are in the atom? _____

d Write down the chemical symbol of nitrogen _____

e The element phosphorus is below nitrogen in the Periodic Table. Phosphorus has an atomic number of 15. Describe **two** ways in which the structure of a phosphorus atom is different from a nitrogen atom.

_____

_____

2. Look at the information about some different atoms:

helium   fluorine   sodium   aluminium

$^{4}_{2}He$ $^{19}_{9}F$ $^{23}_{11}Na$ $^{27}_{13}Al$

a Which atom has the same number of protons and neutrons?

_____

b Which atom is in Group 1 of the Periodic Table? _____

c Which atom has seven electrons in its outer shell? _____

d Which atom has the most neutrons in its nucleus? _____

3. Today the Periodic Table is based on the groups that Dmitri Mendeleev suggested in 1869.

Complete these sentences about his work.

a Mendeleev arranged the elements in order of increasing _____ .

b Mendeleev observed that some elements have similar _____

and _____ properties.

c Mendeleev left gaps in his Periodic Table so that new _____

_____ .

d Why was Mendeleev's way of grouping elements so important?

_____

_____

4. The diagram represents the Periodic Table of elements.

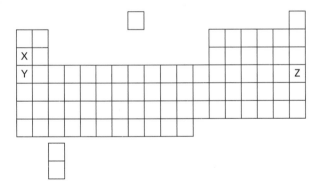

X, **Y** and **Z** represent elements in different positions in the Periodic Table.

a Which letter represents the element with the highest atomic number?

_____

b Elements **X** and **Y** react with cold water to produce hydrogen gas.

    i Describe what you would observe when **X** and **Y** are added separately to cold water.

       _____

       _____

    ii Explain which element will react with water fastest. Give a reason for your answer.

       _____

       _____

    iii Explain why elements **X** and **Y** have similar chemical properties.

       _____

       _____

6.1 Chemical bonds

You will learn:

- To understand that molecules are formed by covalent bonds
- To describe how a covalent bond is formed
- To describe how negatively charged and positively charged ions are formed from atoms
- To describe an ionic bond
- To describe the uses, strengths and limitations of models and analogies

1. Identify the particle diagram that shows **one** type of molecule.

Tick **one** box.

☐ ☐ ☐ ☐

2. Complete the sentences using words from the box.

Ionic	**attraction**	**atoms**	**positively**	**repulsion**
negatively	**neutral**	**ions**	**particles**	**covalent**

The _____ in a molecule are joined together by a _____ bond.

A covalent bond is the electrostatic _____ between the _____ charged

nucleus of both atoms and the shared _____ charged electrons.

3. True or false?

Tick **one** box.

a Atoms are neutral. ☐ True ☐ False

Give a reason for your answer.

_____

_____

b Metals form negative ions. ☐ True ☐ False

Give a reason for your answer.

_____

_____

c The ionic bond is the transfer of electrons between atoms to form positively and negatively charged ions. ☐ True ☐ False

Give a reason for your answer.

_____

_____

4. Label the diagram.

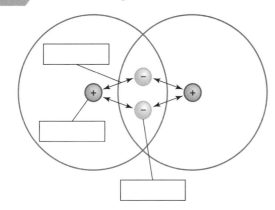

5. Molecules can be represented by dot-and-cross diagrams.

a Draw a dot-and-cross diagram to represent F_2.

Draw the outer shell of each atom in the molecule.

Use dots to represent the electrons on one diagram and x on the other one.

F is in Group 7 of the periodic table, so it has 7 electrons in its outer shell.

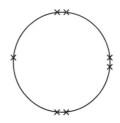

Next draw the molecule by overlapping the two atoms.

Put one electron from each atom in the overlap, so that there are 8 electrons in each outer shell.

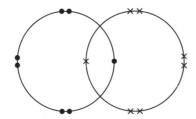

b Draw a dot-and-cross diagram to represent a hydrogen sulfide.

Draw the outer shell of each atom in the molecule. Use dots to represent the electrons in one type of atom and crosses to represent the electrons in the other type of atom.

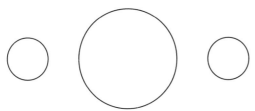

> **Remember**
> to use the periodic table to work out number of electrons in the outer shell

Draw the molecule by overlapping the atoms. Put one electron from each atom in the overlap.

> **Remember**
> In this molecule there will be a covalent bond between each hydrogen atom and the sulfur atom.

c Describe the strengths and limitations of using the dot-and-cross diagram to represent a molecule.

_____

_____

6. Draw a dot-and-cross diagram to represent the carbon dioxide molecule.

7. **a** Complete the word equation:

sodium + chlorine → _____

b The diagram shows the changes that take place during the reaction.

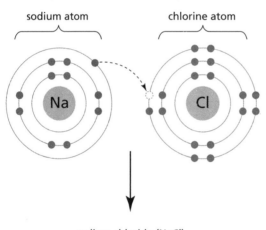

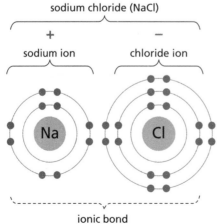

Use the diagram to explain what happens when sodium and chlorine react.

_____

_____

_____

_____

8. An orange flame is seen when sodium metal burns in oxygen, forming sodium oxide.

Challenge Describe the changes that take place to the atoms during the reaction.

You may wish to include a diagram.

_____

_____

_____

_____

6.2 Simple and giant structures

You will learn:

- To understand how the physical properties of an element or compound may be determined by its structure (simple or giant)
- To use observations, measurements, secondary sources of information and keys, to organise and classify organisms, objects, materials or phenomena

1. The diagram shows the structure of four different substances.

Use the words in the box to label each structure.

diamond	oxygen	copper metal	salt crystal	water	plastic

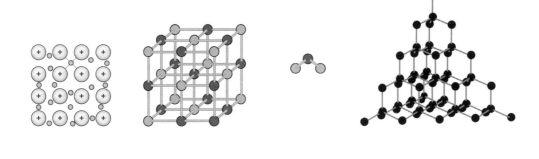

_____ _____ _____ _____

2. Link the statement to the correct explanation.

Draw **one** line from each statement.

The first one has been done for you.

Statement	Explanation
Oxygen is a gas	The electrons are free to move
Iron conducts electricity	It has a giant crystal structure
Diamond is very hard and does not conduct electricity	It has a simple molecular structure
Solid salt cannot conduct electricity but melted salt can	It has a giant molecular structure
Bromine is a liquid at room temperature	

3. Use the words below to complete the sentences.

strong	**ionic**	**covalent**	**metallic**	**molecule**
ion	**forces of attraction**	**repulsion forces**	**weak**	

A sulfur _____ contains eight sulfur atoms. Each atom is held together by a strong

_____ bond. Solid sulfur melts at 115 °C because the _____ between

the sulfur molecules are very _____.,

4. The table shows the properties of some different substances.

Substance	Melting point (°C)	Boiling point (°C)	Can it conduct electricity when solid?	Can it conduct electricity when melted?
A	2852	3600	No	Yes
B	1084	2560	Yes	Yes
C	789	2230	No	No
D	−189	−185	No	No

a Which substance could be copper metal? _____

Give **two** reasons for your answer.

_____

_____

b Which substance could be argon gas? _____

Give **two** reasons for your answer.

_____

_____

c Which substance could be copper chloride? _____

Give **two** reasons for your answer.

_____

_____

6.3 Density

You will learn:

- To describe the mass of a substance in a defined volume as density
- To use the mass and volume of solids, liquids and gases to calculate and compare their densities
- To plan investigations used to test hypotheses
- To use observations, measurements, secondary sources of information and keys, to organise and classify organisms, objects, materials or phenomena
- To evaluate how well the prediction is supported by the evidence collected

1. Write the equation that defines density using the words and units in the list. Write the units in the brackets.

| density | volume | mass | kg | m³ | kg/m³ |

_____ (_____) = _____

(_____) divided by _____ (_____)

2. Thinking about question **1** – what other units we might use for density?

_____

Remember
Always include the units in the answer to every calculation.

3. Carlos has a cube of iron that measures 10 cm along each side. He wants to calculate the density of the iron. Carlos puts the cube on a balance and measures its mass to be 7900 g.

Show Me

a Calculate the volume of the iron cube.

Each side is _____ cm, so the volume is _____ cm × _____

cm × _____ cm = _____

b What is the density of the iron?

_____

_____

_____

c The density of aluminium is 2.7 g/cm³. Suggest why aluminium, not iron, is used to build aircraft.

_____

_____

4. The diagram show how the particles of a substance are arranged in the solid, liquid and gaseous states.

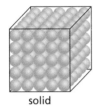 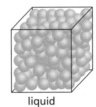 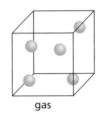

solid   liquid   gas

Use the diagrams to explain why liquids are usually more dense than gases but less dense than solids.

_____

_____

_____

5. The table shows the densities of different materials.

Material	Density in g/cm³	Main properties
Lead	11.3	Strong. Soft. Toxic.
Steel	8.0	Strong. Hard. Non-toxic.
Aluminium	2.7	Strong. Flexible. Non-toxic.
Bamboo	0.3	Strong. Hollow. Can bend a lot.
Fibreglass	2.5	Can be woven to make strong, stiff objects. Can bend a little.
Carbon fibre	1.6	Can be woven to make very strong, stiff objects. Can bend a little.
Nylon	1.1	Can be stretched to make thin fibres. Strong.
Cotton string	1.5	Can be stretched to make thin fibres. Strong.
Polythene	0.9	Can be stretched to make thin fibres. Not very strong.

Look at the diagram. Lead is a metal that was used to make small weights to hold down fishing lines under water.

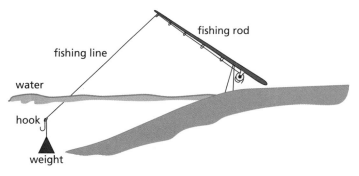

a   Suggest why lead was used to make these small weights.

_____

_____

b   Lead is toxic to animals and people. Fishing weights are now made using other materials. Look at the table. Which material in this table would be best to use instead of lead?

_____

c   Explain your answer to part **b**.

_____

_____

d   Which material would you use to make the fishing line? Explain your answer.

_____

_____

_____

e   Fishing rods can be made from many different materials including bamboo, fibreglass and carbon fibre. Suggest why these materials are suitable for making fishing rods.

_____

_____

_____

6.

Stefan and Petro are looking for the chess set, when they find a box with pieces in from different sets. Stefan picks up the two pawns shown in the diagram.

The pawn on the left feels a lot heavier than the one on the right.

Petro thinks this is because the two pawns have different densities.

The two boys decide to measure the density of both pawns to see if Petro's theory is correct.

a What piece of equipment will the boys use to measure the mass?

_____

b Describe how the boys will measure the volume of each pawn.

_____

_____

_____

_____

Stefan presents the results in a table.

Pawns	Mass	Volume	Density
Left-hand side	68___	25___	_____
Right-hand side	32___	25___	_____

c Add the units of mass, volume and density into the table.

d Calculate the density for each pawn and write your answer in the table.

Show your working.

_____

_____

_____

_____

e Does the data support Stefan's theory? Give a reason for your answer.

_____

_____

_____

7. The density of water is 1.00 g/cm^{-3} at 0 °C.

Challenge The density of ice is 0.92 g/cm^{-3} at 0 °C.

a Explain why this data is unexpected.

_____

_____

_____

b What does this data tells us about the structure of ice and water?

_____

_____

_____

Self-assessment

Tick the column which best describes what you know and what you are able to do.

What you should know:	I don't understand this yet	I need more practice	I understand this
A molecule is formed when two or more atoms join together chemically, through a covalent bond			
A covalent bond is made when a pair of electrons is shared by two non-metal atoms			
An ion is an atom that has gained at least one electron to be negatively charged or lost at least one electron to be positively charged			

	I can't do this yet	I need more practice	I can do this by myself
An ionic bond is an attraction between a positively charged ion and a negatively charged ion			
Some elements or compounds exist as giant structures, which are held together by strong bonds. They have high melting and boiling points			
Some elements and compounds exist as simple structures, which are molecules held together by weak bonds. They have low melting and boiling points			
Density is calculated by mass / volume			
Solids are more dense than liquids and gases because their particles are the closest together			
You should be able to:	**I can't do this yet**	**I need more practice**	**I can do this by myself**
Describe dot-and-cross models and discuss their strengths and limitations			
Classify substances using secondary information			
Plan a range of investigations to obtain appropriate evidence			
Evaluate how well the evidence supports the prediction			

If you have ticked 'I don't understand this yet' or 'I can't do this yet' or mostly 'I need more practice', have another look at the relevant pages in the Student's Book. Then make sure you have completed all the questions in this Workbook chapter and the review questions in the Student's Book. If you have already completed all the questions ask your teacher for help and suggestions on how to progress.

Teacher's comments

End-of-chapter questions

1. **a** The atomic number of magnesium is 12 and the atomic number of oxygen is 8.

Complete the diagrams by adding the correct number of electrons into each shell.

magnesium atom   oxygen atom

b White solid magnesium oxide is produced when magnesium burns in air with a bright light.

Describe how the magnesium and oxide ions are formed.

_____

_____

c What is the structure of magnesium oxide?

Tick **one** box.

☐ A simple molecule

☐ A giant molecular structure

☐ A giant crystal structure

Give a reason for your answer.

_____

_____

2. The diagrams show the giant structure of a metal and a crystal lattice.

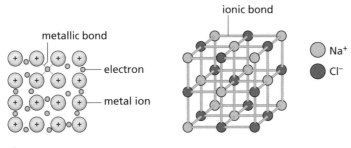

a Complete the sentence:

The metallic bond is the _____ attraction between

_____ ions and electrons.

b Explain why copper is used in electrical cables.

_____

_____

c Name the ions you would expect to find in a crystal of calcium chloride.

_____

_____

d Write down a property you would expect a crystal of calcium chloride to show.

Give a reason for your answer.

Property _____

Reason _____

3. To calculate the density of a material, we need to measure the mass and volume of a sample of the material.

Practical

a An object has straight sides and right-angles at the corners. Describe **one** method for determining the volume of this object that does **not** involve liquids.

_____

_____

b Explain why this method will **not** work for 'irregular' objects (objects with many differently shaped sides or rough surfaces).

_____

_____

The diagram shows how the volume and mass of an irregular object can be found. The markings on the measuring cylinder are in cm³.

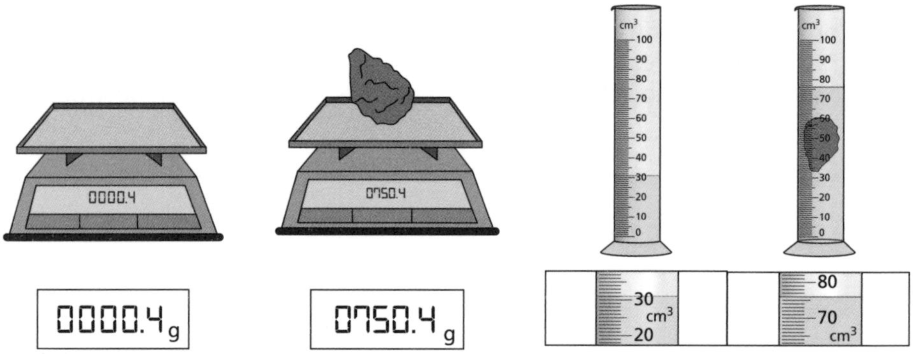

c Write down the readings on the measuring cylinder before and after the object is placed in the water.

_____

_____

d Calculate the volume of the object.

_____

_____

_____

e What is the mass of the object?_____

f Use your answers to parts **d** and **e** to calculate the density of the material the object is made of.

_____

_____

_____

7.1 Changes in chemical reactions

You will learn:

- To understand that mass and energy are conserved in a chemical reaction
- To use scientific knowledge to make predictions
- To evaluate experimental methods, explaining any improvements suggested

1. Complete the sentences.

Choose from the words in the box. Some words may be used more than once or not at all.

products	mass	destroyed	atoms	conservation

The law of _____ of mass states that no _____ are created or

_____ in a chemical reaction. The _____ in the reactants

rearrange to form new products and the total _____ stays the same.

2. Look at the diagram showing two reactants before and after they are mixed together.

Before the reaction

After the reaction

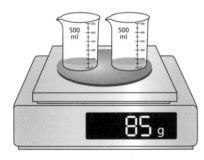

What is the mass after the reaction?

Tick **one** box.

☐ 80 g

☐ 85 g

☐ 90 g

3. Look at the equations for the reaction between magnesium and oxygen.

magnesium   +   oxygen   →   magnesium oxide

a Describe what happens to the atoms during the reaction. _____

b Explain what is meant by the law of conservation of mass.

_____

_____

4. This diagram shows some iron wool on a 'see-saw' balance.

Practical

Explain what will happen to the 'see-saw' when iron wool is heated.

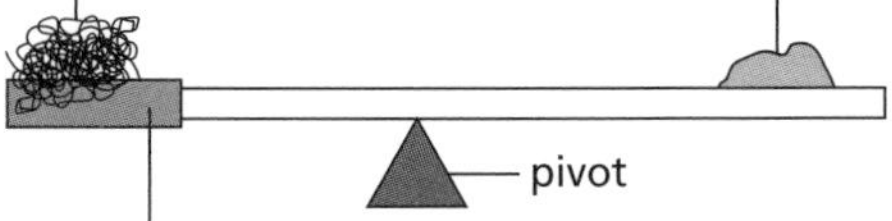

iron wool   weight

Show Me

The left-hand side of the balance will

_____ This is because as the iron   pivot

metal to
protect ruler

wool is heated it _____

Remember

'Explain' means you must say what happens and why it happens.

to form iron oxide, which is _____ than iron.

5. When a candle burns in air, carbon dioxide and water are made.

Practical

The mass of a candle before burning was 15.2 g.

The mass of the candle after 5 minutes of burning was 14.4 g.

a Calculate the rate at which the candle burned in g/min.

_____

_____

b Explain why the mass decreased.

_____

_____

6. **a** Complete the sentence.

The conservation of energy states _____

_____

When zinc is added into copper sulfate solution, an exothermic reaction takes place and the solution becomes warm.

b Suggest reasons why.

_____

_____

_____

_____

7. **Candle investigation**

Mr Yung's class are investigating candles. He sets up the equipment shown in the figure and asks the students to make a prediction.

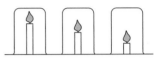

Mingxia thinks the candles will burn for the same time because the beakers are the same size.

Elizaveta thinks the shortest candle will go out first because the carbon dioxide produced will sink to the bottom of the beaker and put the flame out.

Raj thinks that the mass of the candles will all go down by the same amount.

a Explain why Elizaveta thinks that the carbon dioxide will sink.

_____

b Explain why Raj thinks that the mass of the candles will all go down by the same amount.

_____

c Explain why Mr Yung chose beakers that were the same size.

_____

d Each group timed how long it took for the lighted candles to go out.

Here are the results from Mingxia's group.

Candle length in cm	Time candle burned in seconds			
	Experiment 1	Experiment 2	Experiment 3	Mean
10.9	26.0	22.0	18.0	22.0
6.3	22.0	23.0	17.0	
2.4	19.0	25.0	21.0	

Complete the table by working out the mean results. The first one has been done for you.

Here is the working: (26+22+18)/3 = 66/3 = 22

e Write a conclusion by comparing the means of the results.

_____

f Suggest **two** ways you could improve the experiment to test out Raj's prediction.

_____

_____

7.2 Word and symbol equations

You will learn:

- To describe chemical reactions using word or symbol equations (not including balancing)

1. When sodium carbonate is heated, sodium oxide and carbon dioxide gas are formed.

a Name the reactant. _____

b Name the products. _____ and _____

c Complete the word equation for the reaction:

_____ → _____ + _____

2. Look at the word equation given below:

hydrochloric acid + magnesium → magnesium chloride + hydrogen

a Draw a ring around the two products.

b Describe the chemical reaction taking place.

_____

_____

3. Identify the compound that contains the most atoms.

Tick **one** box.

☐ HCl   ☐ Ca(OH)$_2$

☐ H$_2$SO$_4$ ☐ SO$_2$

4. Identify the correct formula for potassium bromide.

Tick **one** box.

☐ KBr^2 ☐ KBr

☐ K$_2$Br   ☐ KBr$_2$

Remember
To work out the formulae of an ionic compounds you will need to balance the charges on the ions so that the overall charge is zero. You can look up the charges on ions in your text book.

5. Explain what is meant by the term 'ion'.

_____

_____

6. Name each of the following compounds from their formulae. You will need to use Table 7.1 in the Student's Book to help you.

a HCl

b H$_2$SO$_4$

c MgCl$_2$

7. Give the formulae for each of the following ions. You will need to use Table 7.1 in the Student's Book to help you.

a Sodium

b Nitrate

c Carbonate

d Calcium

8. Work out the chemical formulae for each of the following compounds. You will need to use Table 7.1 in the Student's Book to help you.

a Sodium nitrate

b Magnesium nitrate

c Sodium oxide

d Calcium bromide

9. Write the symbol equation for this reaction:

calcium carbonate → calcium oxide + carbon dioxide

10. Sulfuric acid and zinc oxide react to form zinc sulfate and water.

a Write the word equation for the reaction.

b Write the symbol equation for the reaction. You will need to use Table 7.1 in the Student's Book to help you.

7.3 Methods for making salts

You will learn:

- To describe the methods used to prepare and purify salts from the reactions between acids and metals, acids and metal carbonates
- To identify hazards and ways to minimise risks
- To choose the appropriate equipment for an investigation and use it correctly
- To evaluate experimental methods, explaining any improvements suggested

1. Which word equation describes the reaction between an acid and a carbonate?

Tick **one** box.

- [ ] acid + carbonate → salt + water
- [ ] acid + carbonate → salt + hydrogen
- [ ] acid + carbonate → salt + water + carbon dioxide
- [ ] acid + carbonate → salt + water + hydrogen

2. Name the salt made when magnesium reacts with hydrochloric acid.

Tick **one** box.

- [ ] Magnesium chloride
- [ ] Magnesium hydrochloric
- [ ] Hydrogen gas
- [ ] Magnesium sulfate

3. Describe how the pH changes as sodium carbonate is added to some sulfuric acid.

Show Me

At the start of the reaction the pH is

_____ because there is only sulfuric

acid in the beaker. When all the sulfuric acid has

reacted, the pH is _____ because the

_____ and water are neutral. If more sodium carbonate

is added the pH will _____ until it reaches 10 or 11.

Remember
Break down the problem into small parts or stages. Then describe each stage in turn.

4. The final stage in making a salt is crystallisation. Explain what 'crystallisation' means.

Worked Example

Crystallisation is a separation technique that is used to separate a solid from a solution. The liquid evaporates leaving crystals of the solid behind.

Remember
If there are two marks for a written question like this then you must make sure your answer has enough detail to get both marks.

5. Read the safety information about calcium.

Practical

Show Me

- Calcium reacts readily with water (or acids) to produce hydrogen, an extremely flammable gas.

- Contact with moisture forms calcium oxide or calcium hydroxide which can irritate the eyes and skin.

- It burns vigorously, but is difficult to ignite.

Describe the safety precautions you should take when making a salt using calcium metal.

Keep it away from naked flames because one of the substances involved

is _____.

Wear _____ to protect your eyes from splashes,

which are _____ .

Do not touch calcium with your hands – wear_____ .

6.

Hassan was making potassium sulfate by adding potassium carbonate powder to some sulfuric acid. During the final stages of the reaction, the potassium carbonate did not fizz when it was added. When Hassan measured the pH he found it was 10.5.

Challenge Explain his result.

_____

_____

_____

_____

7.

Making magnesium sulfate

Practical Oliver and Gabriella want to make some magnesium sulfate. They have been given the following method.

- Add excess metal to 25 cm³ of dilute acid until no more dissolves.

- Filter off the excess metal.

- Evaporate until some solid appears.

- Leave to cool.

- Filter.

a Name the starting materials they should use.

_____

b The diagram shows four pieces of apparatus that can be used to measure volumes of liquids.

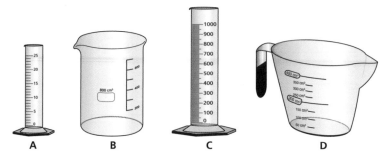

Name the piece of apparatus Oliver and Gabriella should use to measure out the acid. Give a reason for your choice.

_____

_____

c What does 'excess metal' mean?

_____

d Gabriella wants to see the crystals. She wants to continue heating when the solid starts to appear, but Oliver disagrees. He says to get good crystals you must leave the solution to cool. Why is Oliver correct?

_____

_____

e Oliver is about to pour the mixture into the filter funnel when Gabriella stops him because he has forgotten the filter paper.

i Where should the filter paper be?

_____

ii Explain why the mixture needs to be filtered.

_____

_____

f Write a word equation for the reaction.

_____

8. **Making a salt starting with two solutions**

Practical

Blessy and Mia want to make some sodium chloride crystals. The starting materials available are sodium carbonate solution and dilute hydrochloric acid.

a Write a word equation for the reaction.

_____

b Before starting the practical work the girls do a risk assessment. Dilute sodium carbonate solution and dilute hydrochloric acid are both classified as 'low hazard'.

Give **one** safety precaution the girls should take.

_____

c Blessy and Mia are not sure how to carry out the experiment so they do a trial experiment.

- Mia measures out 25 cm³ of hydrochloric acid and pours it into a beaker.
- Blessy then adds 25 cm³ of sodium carbonate solution and observes the reaction.
- They evaporate the mixture until some solid appears before leaving it to cool.

i What does Blessy observe?

_____

ii How do they know when the reaction is over?

_____

d Mia is not happy with this method. She is worried that there might be some unreacted acid in the beaker. Describe a test she could carry out to see if she is right.

_____

_____

e Mia carries out the test and finds that she is right. Describe how the girls should change their method to avoid leaving some unreacted acid at the end of the reaction.

_____

_____

_____

_____

f Mia and Blessy repeat the experiment and this time there is no unreacted acid left at the end of the reaction. Describe the steps the girls need to take to collect the sodium chloride crystals from the reaction mixture in the beaker.

_____

_____

7.4 Displacement reactions

You will learn:

- To identify displacement reactions, and predict their products (involving zinc, magnesium, calcium, iron, copper, gold and silver salts only)
- To plan investigations used to test hypotheses
- To present results appropriately and use them to predict values between the data collected

1. Complete the sentences that follow by choosing from these words:

| more | less | oxidation | neutralisation | displacement |
| non-metal | metal | mixtures | compounds |

A _____ reaction occurs when a _____

reactive metal displaces a less reactive _____ from one of

its _____ .

2. The diagram shows the reactivity series of metals.

a Which metal can be used to displace magnesium from magnesium nitrate?

Tick **one** box.

☐ Iron ☐ Copper

☐ Calcium ☐ Silver

b Which metal can be used to displace iron from iron sulfate?

Tick **one** box.

☐ Iron ☐ Magnesium

☐ Gold ☐ Copper

most reactive

K	potassium
Na	sodium
Ca	calcium
Mg	magnesium
Al	aluminium
C	carbon
Zn	zinc
Fe	iron
Sn	tin
Pb	lead
H	hydrogen
Cu	copper
Ag	silver
Au	gold
Pt	platinum

least reactive

3. The first diagram shows an iron nail being put into some copper sulfate solution. The second diagram shows what it looks like after 20 minutes.

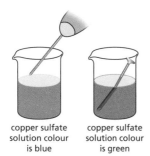

copper sulfate solution colour is blue

copper sulfate solution colour is green

Show Me

a Look at the diagrams and write down your observations.

During the chemical reaction the blue

copper sulfate solution has become

_____ . Part of the

silver-coloured nail has become covered with _____ .

> **Remember**
> An observation is what you *see happening*. You do not need to explain observations.

b A displacement reaction has taken place. Complete the word equation.

iron + copper sulfate → _____ + _____

4. This diagram shows two metals dipped in copper sulfate solution.

How could you tell if zinc or iron is more reactive from this experiment?

_____

_____

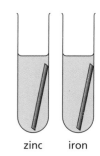

zinc   iron

5. Look at the displacement reactions shown in the two word equations below.

Extension

barium + nickel chloride → barium chloride + nickel

nickel + tin sulfate → nickel sulfate + tin

Put the metals in order of reactivity with the most reactive metal first.

Explain your answer.

_____

_____

_____

6. **Predicting reactivity**

Pierre and Ravjiv are investigating the reactivity of different metals using displacement reactions. This is their method:

- Measure out 10 cm³ of four different metal sulfate solutions into test tubes.
- Add pieces of different metals to the test tubes.
- Carefully observe what happens.
- Record the results.

	Magnesium sulfate	Zinc sulfate	Iron sulfate	Copper sulfate
Zinc	No reaction		Reaction	Reaction
Magnesium		Reaction	Reaction	Reaction
Unknown	No reaction	Reaction	Reaction	Reaction
Iron	No reaction	No reaction		Reaction
Copper	No reaction	No reaction	No reaction	

a Suggest why the boys shaded some parts of the table grey.

_____

b Put the metals in order of reactivity, starting with the most reactive.

_____

c Use your knowledge of the reactivity series to name the unknown metal. Give a reason for your answer.

_____

_____

7.5 Rates of reaction

You will learn:

- To describe and explain how the rate of reaction is affected by changes in concentration, surface area and temperature
- To use scientific knowledge to make predictions
- To represent scientific ideas using recognised symbols or formulae
- To plan investigations used to test hypotheses
- To evaluate how well the prediction is supported by the evidence collected
- To present results appropriately and use them to predict values between the data collected

1. Savita is cooking the dinner. She has forgotten to boil the potatoes. She wants them to cook fast so they are ready on time. Which potatoes will cook the fastest?

Tick **one** box.

- [ ] Whole potatoes
- [ ] Potatoes cut in half
- [ ] Potatoes cut in quarters
- [ ] Potatoes chopped into lots of small pieces

2. Petro is making some biscuits. The first step is to melt some butter in a pan. What could he do to make the butter melt quicker?

Tick **one** box.

- [ ] Put the pan in the fridge
- [ ] Heat the pan
- [ ] Spread the butter on the bottom of the pan
- [ ] Add some water

3. Draw **three** lines between the reacting particles and their concentration.

Particles

Concentration

High concentration

Medium concentration

Low concentration

4. Complete the sentences that follow by choosing words from the box.

less	slower	faster	particle	small	collision
rate	large	more	reactant	product	

Small pieces of solid, especially powders, have a _____ surface area.

The larger the surface area, the _____ the reaction.

This is because there is _____ chance that the particles of the solid

can collide with other _____ particles.

5. Marcus was investigating the rate of reaction between calcium carbonate and hydrochloric acid. He changed the particle size of the calcium carbonate and kept all other variables the same. Marcus timed how long it took for the reaction to take place.

Draw lines to join each particle size to the appropriate reaction time.

Particle size

Single, large chip

Powder

Small chips

Medium chips

Reaction time

1.0 minute

4.2 minutes

8.4 minutes

6.5 minutes

6. Use ideas about particle theory to explain why increasing the temperature of a solution increases the rate of reaction.

Worked Example For a reaction to occur particles must collide.

Particles in a hot solution have more kinetic energy so they move around more quickly.

This means that they will collide with other particles more frequently.

7. Use ideas about particle theory to explain why increasing the concentration of a solution increases the rate of reaction.

Show Me For a reaction to occur particles must

_____ .

Increasing the concentration means the particles are _____ .

This means that particles will collide with other particles _____ .

8. The diagram shows three glasses of orange squash.

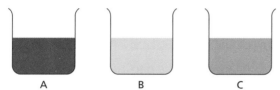

Which glass contains the lowest concentration of orange squash? Give a reason for your answer.

_____

_____

9. Mike and Chen carried out an investigation to see if the concentration of an acid has any effect on the rate of reaction between the acid and magnesium. In their investigation the boys determined the rate of reaction by measuring the time taken for a piece of magnesium to completely dissolve in the acid.

This table shows their results.

Concentration of acid (%)	Time taken to dissolve (s)
100	13
75	22
60	35
50	50
38	100
30	145
25	250

a Plot the points on the axes below and draw a line of best fit.

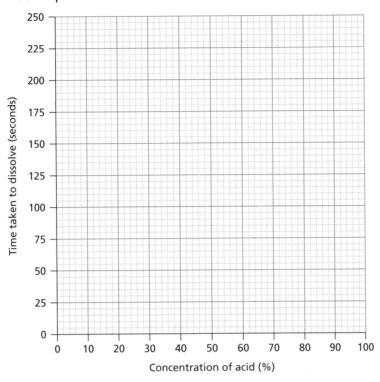

b Use your graph to describe how increasing the concentration of the acid affects the rate of reaction.

_____

c Use your graph to predict the time taken for the reaction if the concentration of the acid was 90%.

_____

d Use your knowledge of the particle theory to explain your answer to part **b**.

_____

_____

_____

10. **Challenge**

Angelique is making a camp fire with her little brother. She asks him to go and find some small pieces of wood to burn. Her brother returns with a large tree trunk.

Explain why Angelique sends her brother away again saying 'Only small pieces of wood will work'.

____

____

____

11. **Practical**

Anastasia and Ahmed are doing an investigation to find how temperature affects the rate of reaction.

This is their method:

- Set up the apparatus shown in the diagram.
- Add 25 cm^3 of hydrochloric acid to the conical flask and record the temperature.
- Add 5 g of calcium carbonate, quickly replace the stopper and record how long it takes to produce 100 cm^3 of carbon dioxide gas.
- Repeat at a different temperature.

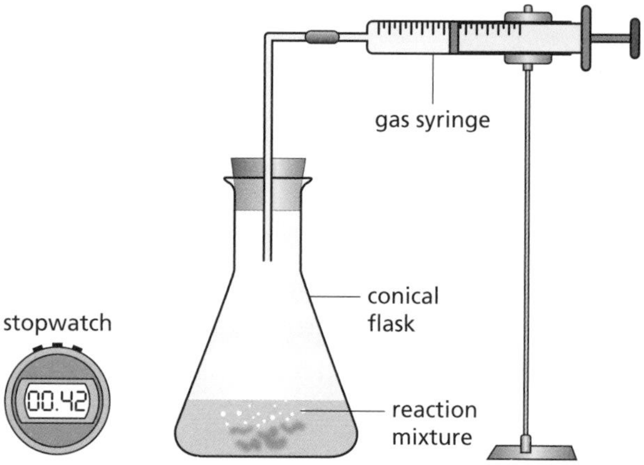

gas syringe
stopwatch
00.42
conical flask
reaction mixture

a Write a word equation for the reaction.

__________________ → __________________ + __________________

b Describe how Anastasia and Ahmed could change the temperature of the reactants.

____

____

c **i** What is the dependent variable?

____

ii What is the independent variable?

____

d   Name **two** variables that should be controlled in order to ensure it is a fair test.

_____

_____

Self-assessment

Tick the column which best describes what you know and what you are able to do.

What you should know:	I don't understand this yet	I need more practice	I understand this
The mass of products in a chemical reaction equals the mass of reactants			
Energy is not created or destroyed; in a chemical reaction it is transferred from one store to another			
A word equation shows us the reactants and products in a reaction			
A symbol equation shows the formulae of the reactants and products			
A salt is formed when metals and metal carbonates react with acids			
The type of salt made in the reaction depends on the acid used			
Salts can be purified using filtration, evaporation and crystallisation			
A more reactive metal will displace a less reactive metal from its compounds			
Displacement reactions can be used to compare the reactivity of different metals			
The rate of a reaction increases as temperature increases: when the concentration of reactant particles increases; and when the surface area of the reactants increases			
Chemical reactions only occur when the reactant particles collide with each other			

	I can't do this yet	I need more practice	I can do this by myself
An increase in temperature increases the rate of a reaction because reactant particles collide more often and with more energy			
An increase in the concentration of a soluble reactant, or in the surface area of a solid reactant, increases the rate of reaction because reactant particles collide more often			
You should be able to:	**I can't do this yet**	**I need more practice**	**I can do this by myself**
Use symbols and formulae to represent scientific ideas			
Plan investigations used to test hypotheses			
Make risk assessments for practical work to identify and control risks			
Interpret results and form conclusions using scientific knowledge and understanding			
Choose the appropriate equipment for an investigation			
Identify control, dependent and independent variables			
Evaluate experimental methods, explaining any improvements suggested			
Make predictions based on scientific understanding and evaluate these against evidence			
Present results and use them to make predictions			

If you have ticked 'I don't understand this yet' or 'I can't do this yet' or mostly 'I need more practice', have another look at the relevant pages in the Student's Book. Then make sure you have completed all the questions in this Workbook chapter and the review questions in the Student's Book. If you have already completed all the questions ask your teacher for help and suggestions on how to progress.

Teacher's comments

End-of-chapter questions

1. Look at the diagram of the reactivity series. It shows some metals in order of reactivity.

a When calcium is mixed with zinc nitrate, a chemical reaction takes place producing calcium nitrate and zinc.

Write the word equation for the displacement reaction.

_____ + _____ →

_____ + _____

most reactive

least reactive

K	potassium
Na	sodium
Ca	calcium
Mg	magnesium
Al	aluminium
C	carbon
Zn	zinc
Fe	iron
Sn	tin
Pb	lead
H	hydrogen
Cu	copper
Ag	silver
Au	gold
Pt	platinum

b Use the reactivity series to predict which of the displacement reactions below will take place.

Tick **one** box.

☐ calcium + sodium chloride

☐ magnesium + iron sulfate

☐ lead + zinc chloride

☐ copper + tin sulfate

c Write a word equation for your answer to part **b**.

_____ + _____ → _____ + _____

d Balance the symbol equation.

_____Na + $ZnCl_2$ → _____ NaCl + Zn

2. Salts can be prepared by the reaction of acids with metals.

a **i** Complete this equation:

acid + metal → salt + _____

ii Circle the word that best describes the reaction.

oxidation	combustion	neutralisation	displacement

b Sodium sulfate is an important salt used to make detergents.

The table lists some substances. Tick the name of the acid used to make sodium sulfate.

Hydrochloric acid	
Sulfuric acid	
Nitric acid	

c Sodium sulfate can be made by the reaction of an acid and carbonate.

Complete the word equation for this reaction.

_____ carbonate + _____ acid →

sodium sulfate + _____ + _____

3. Youssef is making zinc chloride. He uses this method:

Practical

- Add one spatula of zinc to some hydrochloric acid and stir.
- Keep adding zinc and stirring until there is an excess.
- Remove the excess zinc.
- Crystallise the salt solution.

a State **one** hazard when making the salt.

_____

b Describe **one** way of reducing the risk of harm.

_____

c When will Youssef know it is time to stop adding the zinc?

_____

d In the space provided below, draw a labelled diagram to show how the excess zinc is removed.

e Write a word equation for the reaction.

_____

4. Acids and bases are commonly found around the home.

Practical **a** Some indigestion tablets contain calcium carbonate which neutralises excess stomach acid (hydrochloric acid).

i Complete the word equation for the reaction.

calcium carbonate + hydrochloric acid → _____ +

_____ + _____

ii How does the pH change in the stomach after taking these tablets?

_____

b Ammonium sulfate is a salt that is used to make fertilisers. Some people use ammonium sulfate in their garden.

Safia and Lily were making ammonium sulfate in the laboratory. Here is their method:

- Pour 100 cm³ of sulfuric acid into a beaker.
- Add some indicator.
- Add some ammonium hydroxide until the solution is neutral.
- Filter.
- Evaporate and cool.
- Filter.

i Name the piece of apparatus the girls should use to measure out the sulfuric acid.

_____

ii Explain why Safia and Lily added some indicator to the acid.

_____

iii Why did the girls filter the mixture?

_____

_____

iv Describe how Lily and Safia evaporated the final solution. You can include a labelled diagram as part of your answer if it helps.

_____

_____

_____

_____

5. Mary and Mia are making lemonade. They add some sugar to make it sweeter.

Mary adds three sugar cubes to her lemonade.

Mia adds the same mass of sugar granules to her lemonade.

a Whose sugar will dissolve first? Give a reason for your answer.

_____

_____

_____

b What could Mia do to speed up the rate at which her sugar dissolves?

_____

_____

c Use your ideas about particles to explain your answer to part **b**.

_____

_____

_____

_____

Physics

8.1 Energy conservation

You will learn:

- To understand that energy cannot be created or destroyed

1. Objects store energy in two ways.

Complete the sentences using words from the list.

| kinetic | temperature | position | thermal | potential | movement |

A   B

Height

object

moving
object

A   An object stores energy because of its _____ .

This is called _____ energy.

B   An object stores energy because of its_____ .

This is called _____ energy.

2. Vultures are birds that can fly high in the sky without flapping their wings. This is called soaring.

Which word describes the potential energy of a vulture as it rises? Choose the **best** description.

☐ Decreases

☐ Stays the same

☐ Increases

3. The diagram shows a simple pendulum.

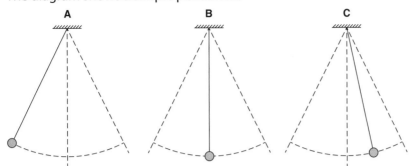

In **A**, the pendulum is at the highest point of its swing. For a very short time, it is not moving.

In **B**, the pendulum is at the lowest point of its swing. At this time, it is moving with its fastest speed.

In **C**, the pendulum is half-way between the lowest and highest points of its swing.

Complete the sentences. Choose the **best** words from the list. You may need to use some words more than once.

minimum	maximum	about the same

a At **A**, the pendulum has _____ kinetic energy and _____ potential energy.

At **B**, the pendulum has _____ kinetic energy and _____ potential energy.

At **C**, the pendulum has kinetic energy and potential energy that are _____ .

b Complete the equation for the pendulum using the mathematical symbols from the list. Choose the **best** symbols from the list.

> **Remember**
> Total energy is always conserved.

+	−	×	÷	=

kinetic energy _____ potential energy _____ constant

4. I am a quantity that can be measured or calculated. I am always there when any activity takes place. I am stored by objects. I can be transferred between objects. I can never be created or destroyed.

What am I? _____

5. Complete the definition of the principle of conservation of energy. Choose the **best** words from the list.

transferred	recycled	destroyed	created	reused

Energy cannot be _____ or

_____ . Energy is _____

from one form to another.

Remember
When energy moves between objects we say energy is transferred. If a quantity is made we say it is created. If a quantity reduces to zero we say it is destroyed.

6. A refrigerator works by using energy from an electrical supply to reduce the temperature inside the storage compartment.

Challenge

Use the principle of conservation of energy to explain why the temperature outside the refrigerator must increase.

_____

_____

_____

_____

7. An electric kettle uses 1.5 kilojoules of energy every second.

a How many joules of energy does it use every second?

b How many kilojoules of energy would the kettle use if it was used for 20 minutes? Include the symbol for kilojoules in your answer.

c How many megajoules of energy is your answer to **b**? Include the symbol for megajoules in your answer.

d Why is it more sensible to write your answer to **b** as kilojoules or megajoules rather than as joules?

8.2 Heating and cooling

You will learn:

- To describe how heat and temperature are different
- To understand how heat dissipation refers to the movement of thermal energy

1. Complete the sentences below to explain the difference between thermal energy and temperature.

Thermal energy is the total amount _____ .

Temperature is the _____ .

The quantity (thermal energy or temperature) that can be measured directly is

_____ .

You can measure this by_____ .

2. Maria's hand feels cold when she touches a block of ice. She says that this is because cold from the ice has gone into her hand. Explain why Maria is wrong.

_____

_____

_____

3. It was once believed that objects that were warm contained more of a substance called 'caloric' than substances that were cooler.

a Name **one** famous scientist whose work helped to replace the 'caloric' theory.

_____

b Give **one** piece of evidence that suggested that the 'caloric 'theory was wrong.

_____

4. A class of students observes a demonstration where a Bunsen burner is used to heat a substance.

Challenge

The amount of gas burned and the temperature of the substance are measured.

From this, the students calculate the amounts of energy transferred.

Chemical energy of gas burned = 2000 J

Thermal energy transferred to substance = 1500 J

One of the students, Blessy, says that energy is not conserved because not all the chemical energy of the gas goes into thermal energy of the substance.

a Blessy's statement is wrong. Explain why she is wrong.

Energy is always _____ .

Blessy has forgotten that _____ .

b Where has the 'missing energy' gone?

_____

5. Whenever energy is transferred from one object to another some thermal energy is produced. Explain why this thermal energy is usually not all useful. Use the word 'dissipate' in your answer.

_____

_____

8.3 Conduction, convection, radiation and evaporation

You will learn:

- To understand how heat dissipation refers to the movement of thermal energy
- To describe the movement of thermal energy during conduction, convection and radiation
- To explain how evaporation causes cooling
- To identify hazards and ways to minimise risks

1. Draw lines to match the names of the heat transfer processes to the descriptions.

Thermal transfer process	Description
conduction	Thermal energy causes a liquid or gas to expand so that hotter parts of the liquid or gas rise, and cooler parts sink
convection	Thermal energy is transferred but does not require matter to travel through
radiation	Thermal energy passes through a substance from particle to particle

2. The three states of matter are solids, liquids and gases.

a List the three states in order so the best conductor is first and the worst conductor is last.

_____ (best)

_____

_____ (worst)

b List the three states in order so the best substance for convection is first and the worst is last.

_____ (best)

_____

_____ (worst)

3. The figure below shows the inside of an electric kettle.

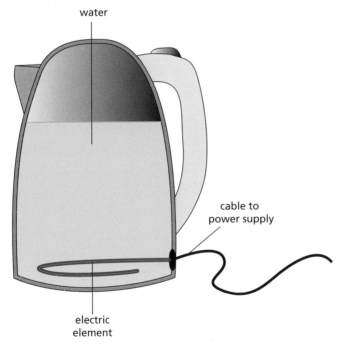

water

cable to power supply

electric element

a What type of material should the electric element be made from? Explain your answer.

_____

_____

_____

b The kettle heats up the water at the bottom. Name the process by which the hotter water moves around the kettle.

_____

c Draw arrows on the diagram to show the directions in which the water moves as it heats up.

4. Modern office buildings use large amounts of glass to make the rooms more brightly lit. The glass has a special coating so that the outer surface is shiny but still allows visible light through the glass.

a A room with a large amount of glass can get very hot. What is the process by which thermal energy enters the room?

_____

b State the source of this thermal energy.

_____

c Explain why the outer surface of modern office glass has a special coating.

_____

_____

5. The figure below shows a pan used for cooking.

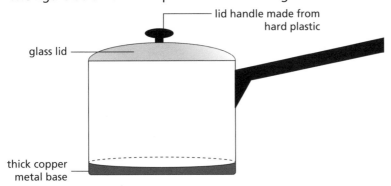

Explain the three labelled features of the pan in terms of how they affect thermal energy transfer.

_____

_____

_____

_____

_____

_____

6. Look at the diagrams below which show three experiments involving thermal transfer.

1
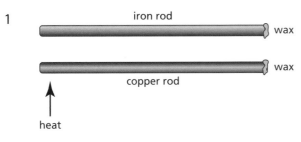

2
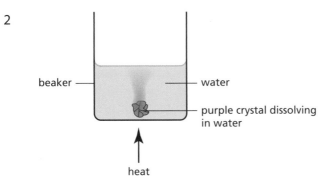

3

heat lamp

square of shiny silver metal

square of matt black metal

Draw a table for each experiment that shows:

a the name of the thermal transfer process that the experiment is designed to show

b an instruction to make the experiment safer.

7. In hot weather, your skin sweats. It produces water that rests on the surface of your skin.

a Use your knowledge of heat transfers to explain why sweating helps your body to cool down.

_____

_____

_____

b If you direct a stream of air over your skin, it can help your body to cool further still. Explain why this happens.

_____

_____

_____

8. Paulo needs to hang some washing out to dry. He needs it to dry as quickly as possible.

Say whether each of the statements below is true or false:

A It doesn't matter if the clothes are spread out or squashed up. _____

B The clothes will dry faster if it's a windy day. _____

C The clothes will dry faster if the weather is humid. _____

D The clothes will dry faster if the weather is warm. _____

Self-assessment

Tick the column which best describes what you know and what you are able to do.

What you should know:	I don't understand this yet	I need more practice	I understand this
Thermal energy is transferred from a hot objects to colder objects			
Energy is conserved: it is never created or destroyed but just transferred			
The units of energy are joules, kilojoules and megajoules			
The amount of energy that has been wasted can be calculated from the difference between the total energy input and the useful outputs of energy			
Temperature is a measure of the average energy each particle in an object has			
Thermal energy can be transferred by conduction, convection, radiation and evaporation			
Thermal energy is not easily transferred through thermal insulators			
Conduction occurs in solids and happens when heat is passed from particle to particle			
Convection occurs in fluids and happens when the particles move around because of having more energy			
Radiation is caused by heat being transferred using waves and can occur in a vacuum			
Silver, shiny objects reflect heat well; black, matt objects absorb and emit heat well			

You should be able to:	I can't do this yet	I need more practice	I can do this by myself
Identify hazards and identify ways to minimise risks			
Use the particle model to explain how conduction and convection works and how evaporation cools things			

If you have ticked 'I don't understand this yet' or 'I can't do this yet' or mostly 'I need more practice', have another look at the relevant pages in the Student's Book. Then make sure you have completed all the questions in this Workbook chapter and the review questions in the Student's Book. If you have already completed all the questions ask your teacher for help and suggestions on how to progress.

Teacher's comments

End-of-chapter questions

1. The diagram below shows a cross-section through a vacuum flask.

   a Use this phrase list to complete the labels in the diagram.

     gap (vacuum) between surfaces

     thick plastic stopper

     shiny metal wall

     plastic supports

b Use the phrase list to add the *main* reason each labelled part in the diagram is used. Each answer may be used once, more than once or not at all.

to reduce radiation

to reduce conduction

to reduce convection

c Vacuum flasks are used to keep the thermal energy in hot drinks high for as long as possible. Predict whether or not a vacuum flask can also be used to keep the thermal energy in cold drinks low. Explain your answer.

_____

_____

_____

2. Explain what is meant by the term 'heat dissipation'.

_____

_____

_____

3. Use the particle model to explain how thermal energy is transferred from one end of a metal spoon to the other during the process of conduction.

_____

_____

_____

4. Explain how hot liquids cool down through the process of evaporation.

_____

_____

_____

5. Give **two** differences between evaporation and boiling.

_____

_____

9.1 Floating and sinking

You will learn:

- To explain how an object floats or sinks using density
- To use scientific knowledge to make predictions
- To represent scientific ideas using recognised symbols or formulae
- To plan investigations used to test hypotheses

1. Use some of the words in the box to fill in the gaps and complete the sentences below.

downwards	**upwards**	**pressure**	**upthrust**	**greater**
	smaller	**deeper**	**shallower**	

Things float in water because of a force called _____ .

This force is caused because the _____ exerted by water increases

as the water gets _____ . This causes an unbalanced force pushing

the object _____ . If this force is _____ than

the force of gravity the object will float.

2. Tia puts a piece of wood into a stream. The wood sinks. What does this tell you about the size of the upthrust compared with the weight of the piece of wood?

_____

_____

3. Draw a force diagram for a log floating in water. Use the diagram to explain why it floats.

Challenge

4. Si uses a forcemeter to weigh a metal block. He then places the metal block into a beaker of water (the block is still attached to the forcemeter). The block does not touch the bottom of the beaker.

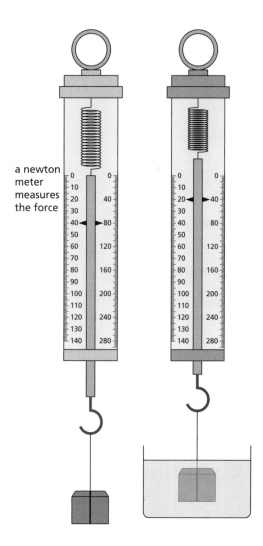

a newton meter measures the force

Which **two** of the statements below are true?

A The block always weighs less when it is suspended in water.

B The block weighs the same in or out of water.

C The block will only weigh less when it is in the water if it floats.

D Upthrust has no effect if the block does not float.

E Upthrust makes the block weigh less when it is suspended in water.

> **Remember**
> Make sure you choose the correct number of options; no more and no fewer.

5. Which **one** of the equations below is correct?

$$density = \frac{volume}{mass}$$

$$density = \frac{mass}{volume}$$

$$density = mass \times volume$$

6. Which **one** of the equations below are correct?

Challenge

$$density = mass \times volume$$

$$volume = \frac{density}{mass}$$

$$mass = density \times volume$$

7. What **two** units is density usually measured in?

_____   _____

8. Water has a density of 1 g/cm³. Which **two** of the objects in the table will float in water?

Object	Density (g/cm³)
Hardwood ruler	1.6
Ice cube	0.92
Steel ruler	8.2
Polystyrene ball	0.2

9. **a** A piece of wood sinks when put into a container of water. What does that tell you about the density of the wood?

_____

b The same piece of wood is now hollowed out so it is in the shape of a boat. Explain why the wood will now float in water.

_____

_____

_____

10. Mercury is a metal that is liquid at room temperature. It has a density of 13.5 g/cm^3. Water has a density of 1 g/cm^3 at room temperature.

What would happen if you tried to mix water and mercury? Explain your answer.

_____

_____

_____

_____

_____

11.

Practical

Nuria thinks salt water is denser than fresh water. She wants to design an experiment to test this hypothesis. She has a small flat piece of wood that floats on water, several small masses, salt, water and a large beaker.

a Describe how Nuria can use this equipment to test her hypothesis.

_____

_____

_____

_____

_____

_____

_____

_____

_____

b What result would you expect Nuria to get if her hypothesis is correct?

_____

_____

12.

Challenge

Water in a beaker is being heated with a Bunsen burner. The water at the bottom of the beaker gets warm first and rises to the top of the beaker. Cold water from the top of the beaker falls to take its place.

a What does this tell you about the density of warm water and cold water?

_____

_____

_____

_____

b What does this tell you about the separation of the water particles in warm water and cold water?

_____

_____

_____

Self-assessment

Tick the column which best describes what you know and what you are able to do.

What you should know:	I don't understand this yet	I need more practice	I understand this
Objects float because of upthrust			
Upthrust is caused by differences in water pressure on the top and bottom of objects suspended in liquids			
Upthrust works against the force of gravity acting on an object			
Objects weigh less when they are suspended in a liquid			
Objects will float on a liquid if they are less dense than the liquid			
Density = mass/volume			

You should be able to:	I can't do this yet	I need more practice	I can do this by myself
Use symbols and formulae to represent scientific ideas			
Use scientific knowledge to make predictions			
Plan an investigation to test a hypothesis			
Interpret results and draw conclusions			

If you have ticked 'I don't understand this yet' or 'I can't do this yet' or mostly 'I need more practice', have another look at the relevant pages in the Student's Book. Then make sure you have completed all the questions in this Workbook chapter and the review questions in the Student's Book. If you have already completed all the questions ask your teacher for help and suggestions on how to progress.

Teacher's comments

End-of-chapter questions

1. **a** A plastic cube has a mass of 240 g. Kris suspends the cube from a forcemeter and sees that it weighs 2.4 N. Kris decides to immerse the cube in water. It is still attached to the forcemeter. What will happen to the forcemeter reading when the cube is suspended in water? Explain your answer.

_____

_____

_____

_____

b The plastic cube has a volume of 100 cm³. Calculate its density. Give your answer in g/cm³.

_____

_____

c Would this plastic cube float in fresh water? (Fresh water has a density of 1 g/cm³). Give a reason for your answer.

_____

_____

d Kris makes a very large hole in the plastic cube which makes it hollow. Explain why the cube can now float on water.

_____

_____

_____

e Kris puts the hollow cube into a bowl of salt water. He notices that the cube floats higher in the salt water than it did in fresh water. What does this tell you about the density of salt water?

_____

10.1 Voltage and resistance

You will learn:

- To understand how to draw circuit diagrams using conventional symbols for the components
- To understand how current and voltage can be measured in series and parallel circuits
- To describe how the addition of cells and lamps affects the measurement of voltage
- To understand how to calculate resistance and describe the effect of resistance on current
- To represent scientific ideas using recognised symbols or formulae
- To choose the appropriate equipment for an investigation and use it correctly

1. Complete these sentences by choosing the correct words. Circle the **best** answer.

a For electrons to move around a circuit, **energy / current** must be supplied.

b We can measure the amount of (the answer from part **a**) at a point in a circuit. We call this measurement the **current / voltage**.

c We can use **an ammeter / a voltmeter** to measure this quantity.

2. **a** What is the maximum output voltage of four 1.5 V cells connected together?

_____V

b What would happen to the output voltage if one of the four cells were connected the wrong way round?

_____

3. Complete the sentences below:

The resistance of a circuit measures how difficult it is for _____ to flow.

The bigger the resistance, the more energy or _____ is needed to make a current flow.

Resistance is measured in units called _____ whose symbol is

the Greek letter _____ .

4. Say whether each statement below is true or false.

A Voltage is a measure of how much charge flows around a circuit._____

B Voltage is a measure of how much energy is supplied to the charges moving

around a circuit. _____

C Voltage is measured in amps. _____

D Most components in electrical circuits have some resistance. _____

E The resistance of a variable resistor can be changed. _____

F Resistance is measured in ohms _____

G A thermistor is a resistor whose resistance changes with different light levels.

_____

H If the resistance of a circuit is increased, the current through the circuit will

get smaller._____

5. Which **one** of the equations below is correct?

A resistance = $\dfrac{\text{current}}{\text{voltage}}$

B resistance = $\dfrac{\text{voltage}}{\text{current}}$

C resistance = current × voltage

6. A 2 V cell is connected to a 10 Ω resistor. What current flows in the resistor?

_____

_____

7. A 6 V cell is connected to a 20 Ω resistor. What current flows in the resistor?

_____

_____

8. What voltage is needed to make a current of 0.2 A flow through a 40 Ω resistor?

_____

_____

9. A current of 0.5 A flows through a resistor when a voltage of 24 V is applied. What is the resistance of the resistor?

_____

_____

10. Complete these sentences:

A _____ is needed to measure the voltage across a lamp in an electric

circuit. It must be connected in _____ with the lamp.

An _____ is needed to measure the current through a lamp in an

electric circuit. It must be connected in _____ with the lamp.

10.2 Measuring current and voltage in series and parallel circuits

You will learn:

- To understand how to draw circuit diagrams using conventional symbols for the components
- To describe current in series and parallel circuits
- To understand how current and voltage can be measured in series and parallel circuits
- To describe how the addition of cells and lamps affects the measurement of voltage
- To represent scientific ideas using recognised symbols or formulae
- To use scientific knowledge to make predictions
- To plan investigations used to test hypotheses
- To describe trends and patterns shown in a set of results, identifying, and explaining, any anomalous results present
- To interpret results, form conclusions using scientific knowledge and understanding and explain the limitations of those conclusions

The circuits below are used in questions **1** to **6**.

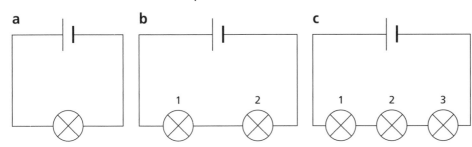

1. The voltage measured across the cell is the same for each circuit. Use ideas about energy to explain why this is.

_____

_____

2. Compare **circuit b** with **circuit a**. Ahmed measures the voltage across the first lamp in each circuit. He finds that the voltage across the first lamp in **circuit b** is about half the voltage across the lamp in **circuit a**.

Explain why this is.

_____

_____

3. We can write the relationship in **circuit a** and **circuit b**:

voltage across lamp 1 in **circuit b** < the voltage across the lamp in **circuit a**

Which of the following statements is true about the voltage across lamp 1 in **circuit c** compared to lamp 1 in **circuit b**? Tick the **best** answer.

☐ voltage across lamp 1 in **circuit c** < voltage across lamp 1 in **circuit b**

☐ voltage across lamp 1 in **circuit c** = voltage across lamp 1 in **circuit b**

☐ voltage across lamp 1 in **circuit c** > voltage across lamp 1 in **circuit b**

4. Ahmed takes **circuit b** apart and makes a new **circuit d**. The two lamps are now connected in parallel with each other and the cell.

a Draw a diagram of **circuit d**. Label each of the lamps as '1' or '2'.

b Which of these statements shows the correct relationship between the voltage of the cell and the voltages across lamps 1 and 2? Tick **one** answer.

☐ V across cell = V lamp 1 = V lamp 2

☐ V across cell = V lamp 1 + V lamp 2

☐ V across cell > V lamp 1 > V lamp 2

☐ V across cell < V lamp 1 < V lamp 2

5. Blessy uses **circuit d** to test the relationship between the voltage across the cell and the voltages across each lamp.

Write a method for Blessy's investigation. Assume that Blessy has only one voltmeter.

_____

_____

_____

6. Blessy extends her investigation to include **circuit b**. She also decides to measure the current in each lamp and the cell. Here is a partly completed table of results.

Challenge

Circuit	Current (A)			Voltage (V)		
	Cell	Lamp 1	Lamp 2	Cell	Lamp 1	Lamp 2
B	3.0	3.0		1.5	0.75	
D	3.0	1.5				

Use your knowledge of current, voltage and also series and parallel circuits to complete the table of results with the values you would expect to observe.

7. Two identical lamps (**X** and **Y**) are connected in parallel with each other. The circuit is supplied by two cells giving a total voltage of 3 V.

a Draw the circuit diagram for this circuit. Label one lamp **X** and the other lamp **Y**.

b State the voltage across each lamp:

$V_X =$ _____ V

$V_Y =$ _____ V

c Each lamp has a resistance of 10 Ω. Calculate the current flowing through lamp **X**.

_____

_____ A

Challenge **d** What is the maximum current flowing in this circuit?

_____ A

8. Jan sets up the circuit shown below:

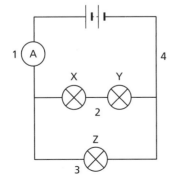

The ammeter measures a current of 0.3 A flowing at position **1.**

a What will the current be at positions **2**, **3** and **4**?

Position **2:** _____ **A**

Position **3:** _____ **A**

Position **4:** _____ **A**

b Jan wants to know the voltage across each lamp. He knows that each cell supplies a voltage of 1.5 V. Before he measures the voltages he decides to predict what each of the voltages would be.

Predict the voltage across each lamp.

Lamp **X:** $V_X =$ _____ V

Lamp **Y:** $V_Y =$ _____ V

Lamp **Z:** $V_Z =$ _____ V

c Add a voltmeter to the circuit diagram to show where Jan would place a voltmeter to measure the voltage across lamp **Y**.

d Add a switch to the diagram that would operate **lamp Z only**.

e Is it possible to add a switch to this circuit that would operate **lamp X only**? Give a reason for your answer.

_____

_____

_____

Self-assessment

Tick the column which best describes what you know and what you are able to do.

What you should know:	I don't understand this yet	I need more practice	I understand this
Current is measured in amps using an ammeter			
Ammeters are placed in series with the component they are measuring the current through			
The current is the same all round a series circuit			
In a parallel circuit the current splits, but the total current remains the same			
The flow of current can be controlled using different components in a circuit			
Increasing the length of resistance wire in a circuit decreases the current			
Voltage is a measure of the energy supplied to electrons by a cell			
A voltage is needed to make electrons flow round a circuit			
If cells are connected in series, their voltages are added together			
Voltage is measured in volts using a voltmeter			
Voltmeters are placed in parallel with the component they are measuring the current through			
Each arm of a parallel circuit gets the full supply voltage			
If there is more than one component in one arm of a parallel circuit the supply voltage is divided between the components			
Resistance is measured in ohms			
Resistance is calculated using the equation $\text{resistance} = \dfrac{\text{voltage}}{\text{current}}$			

You should be able to:	I can't do this yet	I need more practice	I can do this by myself
Plan methods for investigations to test hypotheses			
Choose the appropriate equipment for an investigation			
Interpret results and form conclusions using scientific knowledge and understanding			
Use recognised symbols to draw representative diagrams of series and parallel circuits			
Use scientific knowledge to make predictions			
Describe trends and patterns shown in a set of results, identifying, and explaining, any anomalous results present			
Evaluate how well the prediction is supported by the evidence collected			

If you have ticked 'I don't understand this yet' or 'I can't do this yet' or mostly 'I need more practice', have another look at the relevant pages in the Student's Book. Then make sure you have completed all the questions in this Workbook chapter and the review questions in the Student's Book. If you have already completed all the questions ask your teacher for help and suggestions on how to progress.

Teacher's comments

End-of-chapter questions

1. The diagrams show four different circuits **a** to **d**.

a

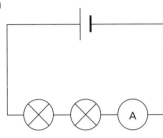

b

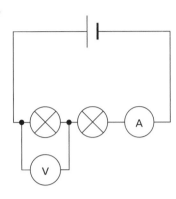

c

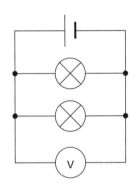

d

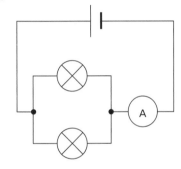

a Which circuit could be used in an investigation to measure **both** current **and** voltage?

_____

b Name the components used to measure:

  i voltage

  _____

  ii current

  _____

c Which circuits include lamps connected in series?

_____

d Which circuits include lamps connected in parallel?

_____

2.

Practical

Angelique investigates the current in a circuit, and how it changes when the length of a resistance wire is changed. Look at the circuit diagram. The symbol for the variable resistor represents the resistance wire.

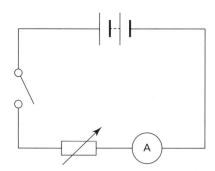

a Predict what you would expect to observe when the length of resistance wire is changed.

_____

_____

Look at the table of Angelique's results.

Length of resistance wire (cm)	Current (A)
10	2.0
20	2.2
30	0.8
40	1.2

b Explain why these results cannot be used to make a clear conclusion about the relationship between the length of resistance wire and current.

_____

_____

c Suggest what further investigations Angelique could do to be able to make a confident conclusion.

_____

_____

3. Draw the circuits for the following investigations or uses.

a Two lamps in parallel with a cell and one switch that controls both lamps.

b Three cells in series with a resistor and a switch.

c This diagram shows two lamps and a light-dependent resistor. Complete the circuit diagram so that the lamps are in parallel with a battery containing two cells, and the light-dependent resistor switches both lamps on and off.

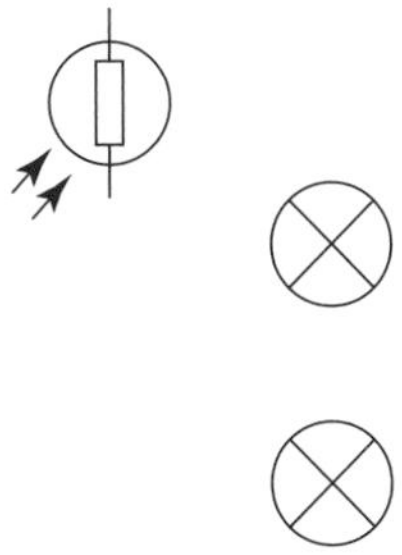

4. Sofia sets up a circuit to investigate the relationship between current and voltage for a fixed resistor.

a Draw a circuit diagram of the circuit Sofia would use for this investigation.

b Use your understanding of circuit electricity to predict what Sofia's results will show.

_____

_____

Challenge **c** Explain the reasons for your prediction.

_____

_____

_____

_____

d Sofia's results are shown below:

Voltage (V)	Current (A)	Resistance (Ω)
0	0	Cannot be calculated
1	0.5	
2	1.0	
3	1.2	
4	2.0	
5	2.5	

i Complete the final column of the table.

ii Which result should Sofia check? Why?

_____

_____

_____

iii Plot Sofia's results on a graph of current (*x*-axis) against voltage (*y*-axis). Draw a line of best fit.

Challenge

iv What feature of your graph shows that the resistor always has the same resistance?

_____

_____

_____

11.1 Loudness and pitch

You will learn:

- To understand how to draw and label waveforms and describe the links between loudness and amplitude and pitch and frequency
- To describe trends and patterns shown in a set of results, identifying, and explaining, any anomalous results present

1. The diagram shows how a wave can be represented on an oscilloscope screen.

Complete the diagram using the words from the list. Not all of the words are used.

amplitude   wavelength   frequency   waveform

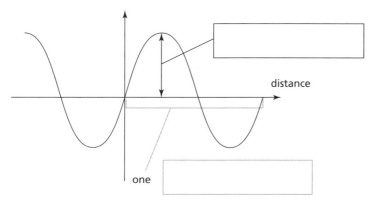

distance

one

2. Look at the table. The key words **1** to **4** match up with the descriptions **A** to **D**.

For each key word, choose the **best** description. Write in a description letter next to each key word.

Key		Description
1 Amplitude		**A** The length of one complete wave
2 Wavelength		**B** The number of waves per second
3 Frequency		**C** The unit of frequency
4 Hertz		**D** The maximum height of a wave

3. Look at the diagrams of sound waves shown on an oscilloscope screen. They are all to the same scale. Choose the **best** diagram to answer each of the following questions. The first question has been answered for you.

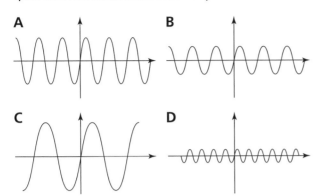

A

B

C

D

Worked Example

a Which is the loudest sound? C

b Which is the quietest sound? _____

c Which sound has the highest frequency? _____

4. The range for human hearing is 20 Hz to 20 000 Hz. Look at the table of sounds. Tick each sound that a human should be able to hear.

	Sound	Frequency (Hz)
☐	Musical note A	440
☐	The call of a bat	45 000
☐	The song of a blue whale	10
☐	Highest note on a piano	4186
☐	Thump of a bass drum	70

5.

Practical

The graph shows the results of a hearing test for Ahmed. The vertical axis shows the minimum loudness Ahmed can hear for a particular frequency of sound. If the sound is quieter than this value, Ahmed cannot hear it.

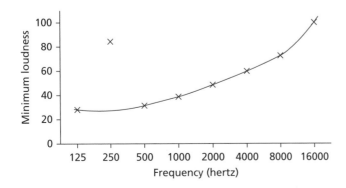

a Describe any pattern you can see in the results. Write your answer in the style of a conclusion.

The graph shows a line that _____.

As the frequency _____.

b Which result does not fit this pattern?

_____

c Ahmed is 68 years old. Suggest how the graph would change for a healthy 18-year-old person.

_____

_____

6. Anastasia is pregnant and has been asked to go to the hospital for an ultrasound scan. Anastasia does not know what ultrasound is. She is worried about the test. Write a short explanation of ultrasound for her. Make sure you include an explanation of why the test will not damage her baby.

_____

_____

_____

7. Explain how sounds can damage our ears.

A loud sound causes the ear drum to vibrate _____.

Our ear drum is a piece of skin tissue that is _____.

This means it is delicate. 'Delicate' means that it _____ easily.

A loud sound can cause this to happen, which means we cannot hear properly until the ear drum is _____.

8. Choose which of these activities should only be carried out if we wear ear protection. Tick **three** boxes to show the **best** choices.

☐ **A** Using an electric drill on a wall   ☐ **B** Playing football

☐ **C** Using a mechanical drill to dig up a road ☐ **D** Going to a rock concert

☐ **E** Driving a car

9.

Challenge

Aiko is a sound technician for a rock band.

The band like to turn up the sound very high and play lots of electric guitar music.

Aiko measured the loudness of one of the concerts. The concert lasted two hours and the loudness changed depending on the song that was being played.

Look at the graph of loudness against time.

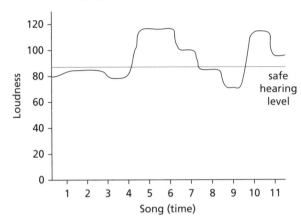

a The horizontal line shows the maximum loudness that is completely safe for humans. Was this concert completely safe for everyone who went to it? Explain your answer.

_____

_____

b Suggest **two** things that Aiko could do to protect her hearing.

1. _____

2. _____

c Explain to Aiko why protecting her hearing is important.

_____

_____

_____

11.2 Interference

You will learn:

- To understand how to represent the interaction or reinforcement of sound waves using waveforms

1. Add the following labels to the diagram below:

crest   **trough**   **amplitude**

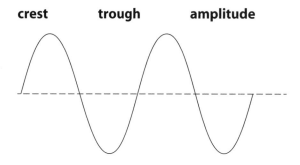

2. In the space below, draw a diagram to show what would happen if two identical waves interfered with each other:

a If crests met crests

b If crests met troughs

c Which of your drawings shows constructive interference? _____

d Which of your drawings shows destructive interference? _____

e Which type of interference is used in noise cancelling technology?

_____

_____

3. Why can't you get good destructive interference if the two original waves are not identical?

_____

_____

_____

4. Draw a pair of waves that would combine to make a wave with 1.5 times the amplitude of the larger of the two original waves.

Challenge

Self-assessment

Tick the column which best describes what you know and what you are able to do.

What you should know:	I don't understand this yet	I need more practice	I understand this
Vibrations produce sound			
The highest point of a wave is called a crest			
The lowest point of a wave is called a trough			
The amplitude of a wave is the height of the wave from its centre to either the crest or the trough			
The frequency of a wave tells you how many vibrations there are each second			
Frequency is measured in hertz (Hz)			
The pitch of a sound depends on the frequency of the sound wave			
The volume of a sound depends on the amplitude of the sound wave			

	I can't do this yet	I need more practice	I can do this by myself
Other animals may have different hearing ranges			
Ultrasound means sound waves which are too high for humans to hear			
Two waves can interfere to make a bigger wave (constructive interference)			
Two waves can interfere to cancel each other out (destructive interference)			
You should be able to:	**I can't do this yet**	**I need more practice**	**I can do this by myself**
Describe trends and patterns shown in results and explain any anomalous results identified			
Understand how an oscilloscope is used to measure sounds			

If you have ticked 'I don't understand this yet' or 'I can't do this yet' or mostly 'I need more practice', have another look at the relevant pages in the Student's Book. Then make sure you have completed all the questions in this Workbook chapter and the review questions in the Student's Book. If you have already completed all the questions ask your teacher for help and suggestions on how to progress.

Teacher's comments

End-of-chapter questions

1. Explain the difference between amplitude and frequency.

_____

_____

2. The diagrams show oscilloscope measurements of four different sounds.

A
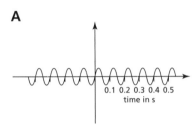

B
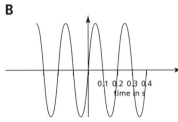

C
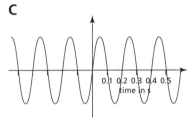

D
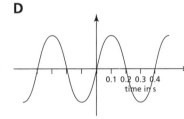

a Which sound is the loudest?

_____

b Which sound has the highest frequency?

_____

c State the unit of frequency.

_____

3. The diagram shows a kettle drum.

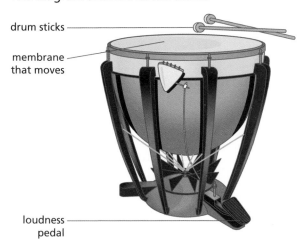

drum sticks

membrane that moves

loudness pedal

Explain how the drum produces a sound.

_____

_____

_____

4. Doctor Strange was visited by a family. Some of them were having problems hearing some sounds.

The doctor decided to test each family member's hearing by measuring the highest frequency sound they could hear. The loudness of each sound was kept the same. The results are shown in this table.

Name	Age	Highest frequency that can be heard (Hz)
Gabriella	74	9000
Carlos	50	16 500
Safia	44	18 000
Lily	24	8000
Ahmed	17	20 000

a Describe any pattern you can see in the results.

_____

_____

_____

b Which result does not fit this pattern?

_____

Earth and space

Chapter 12: Plate tectonics

Chapter 13: Climate change

Chapter 14: Astronomy

12.1 Evidence for plate tectonics

You will learn:

- To explain the evidence for tectonic plate movement, including coastal shapes, volcanoes and earth-quakes, fossil records and the position of magnetic materials
- To describe the uses, strengths and limitations of models and analogies
- To use scientific knowledge to make predictions

1. Choose the best description of the 'jigsaw' evidence for plate tectonics. Tick **one** box.

☐ **A** Two separate continental coastlines that have shapes that appear to fit together.

☐ **B** Two tectonic plates that have collided, one pushing over the other.

☐ **C** A high number of volcanoes that have appeared along the edge of a tectonic plate.

☐ **D** Mountains that have been formed where two plates push together.

2. Which continents that we can see today are thought to have been joined together to form a supercontinent? Tick **one** box.

☐ **A** Just Africa and South America

☐ **B** Just Africa, South America and Australia

☐ **C** Just Asia and Europe

☐ **D** All continents

3. Which of the following natural events would you expect to occur at boundaries between tectonic plates? Tick **three** boxes.

☐ **A** Earthquakes

☐ **B** Hurricanes

☐ **C** Formation of mountains

☐ **D** Volcanoes

☐ **E** Glaciers

4. In which types of rocks are fossils mainly found? Tick **one** box.

☐ **A** Igneous

☐ **B** Sedimentary

☐ **C** Molten

☐ **D** Metamorphic

5. Sort the following sentences into order to describe how fossils can support the theory of tectonic plates. Write the numbers **1** to **5** in the boxes. The first has been done for you.

☐ **A** The animals die and their remains are buried under layers of rock.

☐ **B** People discover fossils of the same animals on two continents that are separated by thousands of kilometres of ocean.

[1] **C** Animals that live on land evolve on the supercontinent.

☐ **D** The supercontinent breaks apart and the continents drift apart from each other.

☐ **E** The weight of the rock layers causes the remains to form fossils.

6. The sentences below describe different types of evidence for the theory of tectonic plates. Sort the sentences into those that describe evidence found on the ocean floor [**OF**] and those that describe evidence found on continental coastlines [**CC**]. Write '**OF**' or '**CC**' in each box. One has been done for you.

[OF] **A** Magnetised basaltic rock produced by molten rock that cools quickly

☐ **B** Sedimentary rock containing fossils

☐ **C** Matching rock layers on coastlines

☐ **D** 'Stripes' of rock magnetised in different directions

☐ **E** Matching shapes of land

7. The picture shows a waterfall in a rift valley in Iceland.

a Use your knowledge of tectonic plates to describe how a rift valley forms.

A rift valley forms at a _____.

Two tectonic plates are _____.

Molten rock rises to the surface and cools, then _____

_____.

b Suggest **two** further types of evidence that could be found in Iceland to support the theory of plate tectonics.

1. _____

2. _____

8. Sketch a diagram to show how rock layers on coastlines provide evidence for the theory of tectonic plates.

9. The diagram shows the plate boundary under the middle of the Atlantic Ocean.

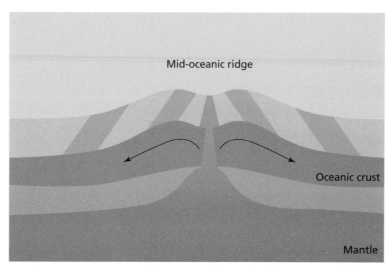

Mid-oceanic ridge

Oceanic crust

Mantle

a   Draw arrows and magnetic pole labels 'S' and 'N' to show how the magnetic field of rocks has been found to vary at the mid-Atlantic ridge.

b   Describe how these magnetic rocks are produced.

_____

_____

c   Explain how these magnetic fields provide evidence supporting the theory of tectonic plates.

_____

_____

_____

10.   How fast does a tectonic plate move? Explain your answer and include a suggestion for how this movement could be measured.

Challenge

_____

_____

_____

_____

_____

_____

_____

12.2 Explaining plate movement

You will learn:

- To explain how convection currents cause tectonic plate movement
- To describe the uses, strengths and limitations of models and analogies
- To identify hazards and ways to minimise risks

1. There are different processes that transfer energy. Choose the process that is involved in the movement of tectonic plates. Tick **one** box.

- [ ] **A** Conduction
- [ ] **B** Convection
- [ ] **C** Radiation
- [ ] **D** All three of the above

2. Complete the sentences using the words from the box.

convection current	**tectonic plates**
conduction layer	**molten rock**

Beneath the Earth's crust is a deep layer of _____ .

This moves because heat from the Earth's core causes a _____ .

The movement causes sideways movements in parts of the crust called _____ .

3. Look at the diagram of coloured liquid in a beaker. At which point should a source of heat be placed to produce the movement of liquid shown by the arrows? Tick **one** box.

- [ ] **A**
- [ ] **B**
- [ ] **C**
- [ ] **D**

4. What is the main source of heat that causes currents in the mantle? Tick **one** box.

☐ **A** The Earth's core

☐ **B** Activity of volcanoes

☐ **C** Radiation from the Sun

☐ **D** Effect of the gravitational force of the Sun on the Earth

Questions **5** to **8** refer to an experiment that uses the following diagram of apparatus.

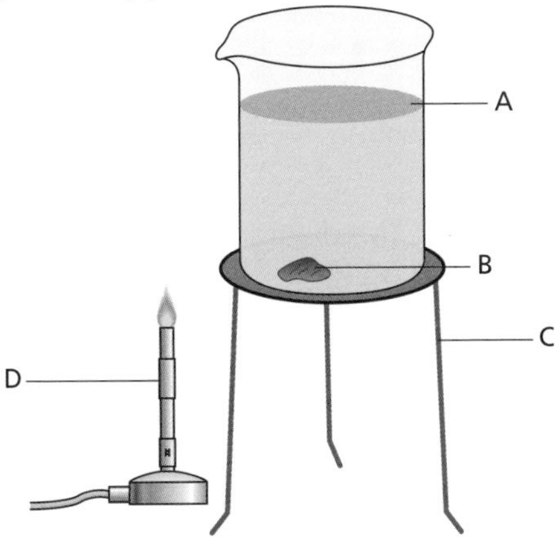

5. Name the pieces of equipment labelled **A** to **D**.

Practical

A _____

B _____

C _____

D _____

6. Briefly describe the method for this experiment.

Practical

A coloured crystal is _____

Show Me

Heat is applied to one corner under _____

The coloured crystal _____

_____

7. Suggest **two** safety precautions that should be used in this experiment.

Practical

1. _____

2. _____

8.

Practical

a State what this apparatus can be used to model. Tick **one** box.

☐ **A** Conduction of heat in metals

☐ **B** Two tectonic plates moving apart

☐ **C** A volcano

☐ **D** Convection currents in the Earth's mantle

b Describe briefly a limitation of this experiment as a model.

_____

_____

9. The diagram shows the boundary between two tectonic plates and the mantle underneath them.

a Describe briefly what is occurring at this plate boundary.

_____

_____

b What effect of the movements at a plate boundary is shown in this diagram?

A Rift valley

B Mid-ocean ridge

C Volcano

D Fold mountains

c Add **at least four** arrows to show how the mantle is moving to cause the movement of the tectonic plates.

10. Scientists have discovered that tidal forces due to the orbit of the Moon also have an effect on the movement of tectonic plates.

Challenge

a Use the effects of forces that produce ocean tides on Earth as a model to suggest how these forces can also affect the Earth's mantle.

_____

_____

_____

_____

b We see effects every day from ocean tides. Suggest **one** limitation with using the ocean tide model to describe how tidal forces affect tectonic plates.

_____

_____

Self-assessment

Tick the column which best describes what you know and what you are able to do.

What you should know:	I don't understand this yet	I need more practice	I understand this
The scientific theory of plate tectonics is supported by a number of types of evidence			
The shapes of the coastlines of continents that are far apart suggest that these continents were once joined together			
Rock layers in coastlines that are now thousands of kilometres apart match up, suggesting these coastlines were once joined together			
Earthquakes and volcanoes occur much more frequently in areas that are boundaries between two tectonic plates, suggesting that the plates are moving			
Fossils of particular species of extinct animals and plants that could not cross oceans have been found in different continents, suggesting that the continents were once joined together			

	I can't do this yet	I need more practice	I can do this by myself
The Earth's magnetic field reverses from time to time, causing different arrangements of magnetic materials in stripes moving outwards from plate boundaries			
Heating a liquid in one area causes particles of the liquid to rise upwards, spread outwards and then fall back down in an area where the liquid cools			
This movement due to heating is called a convection current			
The Earth's core causes heating in parts of the mantle, creating convection currents			
These convection currents are responsible for the movement of tectonic plates			
You should be able to:	**I can't do this yet**	**I need more practice**	**I can do this by myself**
Describe the evidence for plate tectonics including matching coastlines and rock layers, patterns of earthquakes and volcanoes, fossils and magnetic alignment of rocks moving outwards from a plate boundary			
Describe the use, strengths and limitations of models and analogies			
Make predictions based on scientific knowledge of convection currents and plate tectonics			
Identify hazards and suggest ways to minimise risks			

If you have ticked 'I don't understand this yet' or 'I can't do this yet' or mostly 'I need more practice', have another look at the relevant pages in the Student's Book. Then make sure you have completed all the questions in this Workbook chapter and the review questions in the Student's Book. If you have already completed all the questions ask your teacher for help and suggestions on how to progress.

Teacher's comments

End-of-chapter questions

1. Choose the generally accepted explanation for the movement of tectonic plates. Tick **one** box.

☐ **A** Earthquakes cause plates to move apart from each other.

☐ **B** The magnetic field of the Earth causes layers of rock to move in different directions.

☐ **C** Climate change due to human activities.

☐ **D** Convection currents in the molten rock of the mantle pull on the tectonic plates 'floating' above.

2. This question is about the history of the Earth's surface.

a Name the 'supercontinent' from which the continents we know today are thought to have come.

_____

b Describe the process that caused this supercontinent to separate.

_____

_____

c List **three** types of evidence that scientists have found to support this theory.

_____

_____

_____

3.

Practical

Mount Everest is the world's highest mountain. Scientists have been measuring its height since 1849. The 1849 measurement could only be carried out from a long distance away and used devices called giant theodolites which required 12 people to carry each one.

Some of the measurements and the years when they were made are shown in the table.

Year	Height
1849	9200 m
1856	8840 m
1955	8848 m
1975	8848.13 m
2005	8848.43 m

a Suggest why more recent measurements have produced more precise values.

_____

_____

b Was the measurement in 1849 accurate? Explain your answer.

_____

_____

Everest is part of a range of mountains called the Himalayas, formed over the boundary between two tectonic plates.

c Explain how the Himalayas formed.

_____

_____

d Suggest **one** type of evidence that could be searched for on Everest to support this explanation.

_____

_____

e Predict how you would expect the height of Everest to change over the next 1000 years. Justify your answer.

_____

_____

4. The diagram shows three tectonic plates and the mantle underneath them.

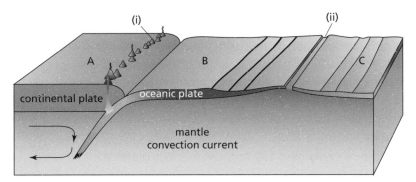

a Add **up to six** arrows to show the direction of movement of molten rock in the mantle under plates **B** and **C**.

b Name the features **(i)** and **(ii)**.

i _____

ii _____

c Describe what is occurring at the plate boundary between plate **A** and plate **B**.

_____

_____

5. Scientific research into the fossil records can be used as evidence to show how continents were once joined together.

Challenge

a Fossils are often found on coastlines with layers of rock. Explain how this use of science can provide evidence for the movement of tectonic plates.

_____

_____

_____

b Fossils are not generally found in areas near mid-ocean ridges. Explain why this is and what other process provides evidence for the movement of tectonic plates in these areas.

_____

_____

_____

13.1 The carbon cycle

You will learn:

- To describe how photosynthesis, respiration, feeding, decomposition and combustion make up the carbon cycle
- To describe trends and patterns shown in a set of results, identifying, and explaining, any anomalous results present
- To interpret results, form conclusions using scientific knowledge and understanding and explain the limitations of those conclusions
- To evaluate experimental methods, explaining any improvements suggested

1. Where in plant cells does photosynthesis take place? Tick **one** box.

☐ **A** Nucleus

☐ **B** Mitochondria

☐ **C** Cytoplasm

☐ **D** Chloroplasts

2. Where in plant cells does respiration take place? Tick **one** box.

☐ **A** Nucleus

☐ **B** Mitochondria

☐ **C** Cytoplasm

☐ **D** Chloroplasts

3. Match the descriptions of processes from the carbon cycle to their names. Draw a line from each process name to the correct description.

respiration	bacteria and fungi feed on dead animals and plants
combustion	plants take in carbon dioxide and produce glucose and oxygen
photosynthesis	plants and animals release energy through the conversion of glucose and oxygen to carbon dioxide and water
decomposition	substances react with oxygen and produce oxides, light and heat

4. **One** of the processes in the carbon cycle depends directly on sunlight. Name the process.

☐ **A** Respiration

☐ **B** Photosynthesis

☐ **C** Combustion

☐ **D** Decomposition

5. Human activities account for most of the carbon dioxide produced by **one** of the processes in the carbon cycle. Name the process.

☐ **A** Respiration

☐ **B** Photosynthesis

☐ **C** Combustion

☐ **D** Decomposition

6.

Practical

A scientist wants to investigate how light affects photosynthesis using pondweed. Sort the variables by their type: are they independent (**I**), dependent (**D**) or control (**C**) variables? Write **one** letter in each box.

☐ Amount (intensity) of light

☐ Size of pondweed sample

☐ Temperature

☐ Number of gas bubbles seen

7.

Practical

A scientist investigates the products of aerobic respiration. Describe tests the scientist can carry out to demonstrate the presence of these products.

_____

_____

8. This is a question about the role of carbon dioxide sinks in climate change.

a Choose the correct description of a carbon dioxide sink. Tick **one** box.

☐ **A** A natural source or producer of carbon dioxide

☐ **B** An artificial source or producer of carbon dioxide

☐ **C** Something that takes carbon dioxide out of the atmosphere

☐ **D** Something that takes oxygen out of the atmosphere

b One of the processes carried out by plants acts as a carbon dioxide sink. Name this process and explain how it works.

_____

_____

_____

9. Human activities can affect several processes in the carbon cycle.

a Describe how human activities affect the amount of carbon dioxide produced by combustion.

_____

_____

b Suggest how humans have affected the amount of carbon dioxide produced by respiration in animals.

_____

_____

10. 'Decomposition is the central process in the carbon cycle.' Do you agree or disagree with this statement? Explain your answer.

Challenge

_____

_____

_____

_____

13.2 Impacts of climate change

You will learn:

- To describe sea level changes, flooding, droughts and extreme weather events as impacts of climate change
- To describe how societies, industries and research use science
- To discuss the global environmental impact of some uses of science

1. An unusually long period with low or no rainfall, causing water shortages.' What does this sentence define? Tick **one** box.

☐ **A** Wild fire

☐ **B** Desert

☐ **C** Drought

☐ **D** Climate change

2. Complete the sentences using the words from the box.

water	oxygen	carbon dioxide

Crops may wilt and die if they lack _____ .

A stray spark or lightning strike may cause the dead plants to combust, a reaction that uses up

_____ .

The burning plants produce the greenhouse gas _____ .

3. What effects are produced by a hurricane? Tick **one** box.

☐ **A** Very high winds and intense rainfall

☐ **B** Very high winds and a heatwave

☐ **C** Intense rainfall and a heatwave

☐ **D** Intense rainfall and a severe cold spell

4. Which of the following is **not** an example of an extreme weather event? Tick **one** box.

☐ **A** A typhoon

☐ **B** Rising average sea levels

☐ **C** Very heavy rainfall

☐ **D** Drought

5. Cities on coasts are more likely to be at risk from the effects of climate change than cities away from coasts. Choose **two** reasons for this. Tick **two** boxes.

☐ **A** More frequent droughts

☐ **B** Rising average sea levels

☐ **C** Extreme cold spells

☐ **D** More intense rainfall

☐ **E** More frequent hurricanes or typhoons

6.

Practical

a A scientist compares the growth of seedlings given different amounts of water. Sort the variables by their type: are they independent (**I**), dependent (**D**) or control (**C**) variables? Write **one** letter in each box.

☐ Height of seedling

☐ Amount of water

☐ Amount of light

☐ Size of pot each seedling is planted in

b Explain why an investigation can only be a fair test if some variables are controlled.

_____

_____

7. Teams of scientists are developing varieties of crops that are more resistant to drought.

a Explain what 'drought resistance' means.

_____

_____

b Suggest why people may need these crops in future.

_____

_____

8. **a** Which of the following events could have been made worse by climate change? Tick **two** boxes.

☐ **A** The eruption of the volcano Mount Pinatubo in June 1991

☐ **B** The tsunami in the Indian Ocean in December 2004

☐ **C** Hurricane Sandy in the Caribbean in 2012

☐ **D** The severe drought and wild fires in Australia in 2019/20

☐ **E** The earthquake in Nepal in 2015

b Suggest why scientists predict there will be more frequent and more extreme weather events in the next 100 years.

_____

_____

_____

Challenge **c** Explain why scientists cannot say for certain that particular events have been affected by climate change.

_____

_____

_____

9. The city of Venice in Italy is built on 100 small islands in a coastal lake. People travel around Venice by a combination of walking, using bridges and using boats as 'taxis'. For normal high tides, the buildings and public areas remain dry. However, if the water level rises 80 cm above a normal high tide, parts of the city become flooded.

a Suggest **two** situations in which a high tide might rise much higher than normal.

_____

_____

If these two situations occur at the same time, larger areas of the city become flooded. In 2019, the whole city flooded because the sea level rose 187 cm above the normal high tide.

b Explain why scientists expect Venice will flood completely more often over the next 100 years.

_____

_____

_____

161

c Suggest **two** ways in which the effects of flooding might be reduced in Venice.

_____

_____

10.

Challenge

Over 97% of climate scientists agree that the Earth's climate has changed due to human activities over the past 200 years. They predict that unless governments in all countries of the world make changes in some human activities, hundreds of millions of people will be affected directly by the effects of climate change.

a Give **three** ways in which these large numbers of people could be affected.

_____

_____

_____

b Explain why all countries of the world need to make changes, even though not all countries may suffer the effects listed in part **a**.

_____

_____

_____

_____

Self-assessment

Tick the column which best describes what you know and what you are able to do.

What you should know:	I don't understand this yet	I need more practice	I understand this
Carbon-containing substances are important to all life on Earth			
Photosynthesis in plants takes in carbon dioxide, water and sunlight and produces oxygen and energy-storing sugars such as glucose			
Respiration is the process by which organisms release energy through the conversion of glucose and oxygen to carbon dioxide and water			
Bacteria and fungi feed on dead animals and plants in the process called decomposition, which also releases carbon-containing substances			

	I can't do this yet	I need more practice	I can do this by myself
Combustion is the reaction of substances with oxygen, releasing energy			
Combustion of carbon-containing substances produces carbon dioxide			
Photosynthesis, respiration, decomposition and combustion all form parts of the carbon cycle			
Climate change is already having measurable effects on sea level and the frequency of extreme weather events			
Extreme weather events include hurricanes and typhoons, long spells of very heavy rainfall leading to flooding and long spells of low rainfall leading to droughts			
Sea levels are rising and will continue to rise as a result of changes in climate			
You should be able to:	**I can't do this yet**	**I need more practice**	**I can do this by myself**
Describe trends and patterns shown in a set of results and identify anomalous results			
Evaluate experimental methods and suggest improvements			
Interpret results and form conclusions using scientific knowledge			
Discuss the global environmental impact of some uses of science			
Discuss the potential global environmental impacts of science and human activities			

If you have ticked 'I don't understand this yet' or 'I can't do this yet' or mostly 'I need more practice', have another look at the relevant pages in the Student's Book. Then make sure you have completed all the questions in this Workbook chapter and the review questions in the Student's Book. If you have already completed all the questions ask your teacher for help and suggestions on how to progress.

End-of-chapter questions

· ·

1. Which processes in the carbon cycle take place inside living plants? Tick **two** boxes.

☐ **A** Combustion

☐ **B** Decomposition

☐ **C** Photosynthesis

☐ **D** Aerobic respiration

2. This question is about flooding.

a State what type of extreme weather event can cause flooding.

_____

_____

b Suggest **two** things that governments can do to reduce the impact of flooding.

_____

_____

c Explain how increasing levels of carbon dioxide in the atmosphere can cause more flooding.

_____

_____

_____

3.

A student investigated the effects of light on the rate of photosynthesis using pondweed. The light intensity increases as the lamp is brought closer to the sample. At each distance, the student counted the number of bubbles produced in one minute before moving the lamp to the next distance.

Result	Distance of lamp from sample (cm)	Number of bubbles counted in 1 minute	Length of pondweed sample (cm)
1	35.0	0	14.0
2	30.0	3	14.0
3	25.0	7	14.0
4	20.0	18	16.5
5	15.0	31	14.0
6	10.0	58	14.0

a **One** of the results should not be used in the analysis of the investigation. State which result and explain why it should not be used.

_____

_____

b Suggest how the student could adjust their method to make their measurements more precise.

_____

_____

c Write a conclusion for the investigation.

_____

_____

_____

4. The diagram shows a cycle.

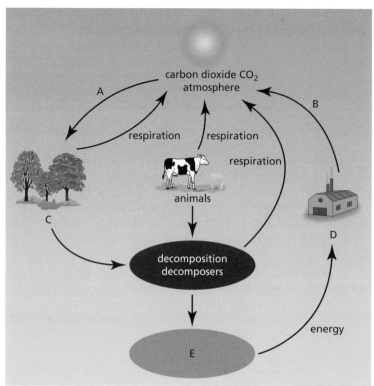

a Name this cycle.

_____

b Name the processes labelled '**A**' and '**B**'.

A _____

B _____

c Describe **three** ways in which the organisms labelled '**C**' are involved in the carbon cycle.

_____

_____

_____

d Suggest **two** examples of human-made objects that could be named in the part of the diagram labelled '**D**'.

_____

_____

e Name the type of substances that appear in section '**E**' and explain how they contribute to climate change.

_____

_____

_____

5.

Challenge

Scientists developing strategies to reduce the effects of climate change often suggest planting large numbers of trees.

Explain how this strategy could reduce the effects of climate change.

_____

_____

_____

14.1 Collisions

You will learn:

- To describe the theory for the Moon formation known as collision theory and its evidence
- To describe how an asteroid colliding with the Earth could cause climate change and mass extinction events
- To understand that models and analogies can change according to scientific evidence
- To use observations, measurements, secondary sources of information and keys, to organise and classify organisms, objects, materials or phenomena
- To evaluate how well the prediction is supported by the evidence collected
- To interpret results, form conclusions using scientific knowledge and understanding and explain the limitations of those conclusions
- To use scientific understanding to evaluate issues

1. When did the Moon form? Tick **one** box.

   ☐ **A** About 100 million years before the rest of the Solar System

   ☐ **B** At about the same time as the rest of the Solar System

   ☐ **C** About 100 million years after the rest of the Solar System

   ☐ **D** About 65 million years ago

2. What is the main evidence for the age of the Moon? Tick **one** box.

   ☐ **A** The orbit of the Moon around Earth

   ☐ **B** Asteroid impact craters on Earth

   ☐ **C** Asteroid impact craters on the Moon

   ☐ **D** Rocks from the Moon

3. Match the names of the hypotheses to the descriptions. Draw a line from each hypothesis to its description.

collision hypothesis	the idea that the Moon and Earth formed together, close to each other, at the same time
capture hypothesis	the idea that a large object, roughly the same size and mass as the planet Mars, collided with the Earth, releasing rocks that were pulled together to form the Moon
co-formation hypothesis	the idea that the Moon is a large asteroid that has been pulled into orbit around the Earth

4. Complete the sentences using the words from the box.

meteor	meteorite	asteroid

A piece of rock found in space that formed around the same time as the planets is a/an

_____.

If this rock enters the atmosphere we call it a/an _____ .

If the rock collides with the Earth's surface, the piece of rock left behind is called a/an _____.

5. The following list describes evidence found about the Moon and how it formed. Sort this evidence into two groups: evidence that suggests the Moon is similar to most asteroids (**A**), and evidence that suggests the Moon is different to most asteroids (**D**). Write **A** or **D** in each box.

☐ The Moon has no atmosphere.

☐ The Moon is very round in shape.

☐ The Moon follows a stable orbit around the Earth.

☐ The Moon has no liquid water on its surface.

6. Describe **two** types of evidence that suggest asteroids have collided with the Earth in the past.

_____

_____

7. Scientists have found evidence that large asteroids have collided with Earth many millions of years ago.

Show Me

a Describe **four** likely effects of a collision of this type.

Melting rocks are thrown _____.

There are intense blasts of _____.

There are shock waves, in which _____.

The Earth's crust _____.

b Suggest why there may be climate change after an asteroid collision.

_____

_____

8. The picture shows the surface of the Moon as seen from Earth.

a There are many circular craters on the Moon (such as the one at the centre bottom of the picture). Explain what has caused these craters.

_____

_____

b Describe how it has been possible for scientists on Earth to analyse samples of rock from the Moon.

_____

_____

c Some rock samples from the Moon are similar to rocks found on Earth. Others are different. Explain what scientists have concluded from this evidence.

_____

_____

_____

9. Should people be concerned about the possible impact of an asteroid on Earth? Explain your answer.

_____

_____

_____

10. From the surface of Earth, we can only ever see the same half of the Moon's surface.

Challenge **a** Describe the relationship between the rotation of the Moon about its axis and the Moon's orbit around Earth that explains this statement.

_____

_____

b What does this relationship suggest about the Moon's orbit? Tick **one** box.

☐ **A** It has changed by large amounts over hundreds of thousands of years.

☐ **B** It has changed by large amounts over hundreds of millions of years.

☐ **C** It has changed by small amounts over hundreds of thousands of years.

☐ **D** It has stayed the same over hundreds of millions of years.

c Use this as evidence to suggest when the Moon formed.

_____

_____

14.2 Observing the Universe

You will learn:

- To understand how stars can form from the clouds of dust and gas known as nebulae
- To use scientific knowledge to make predictions
- To understand that models and analogies can change according to scientific evidence
- To use observations, measurements, secondary sources of information and keys, to organise and classify organisms, objects, materials or phenomena
- To evaluate how well the prediction is supported by the evidence collected
- To discuss the development of scientific knowledge over time because of collective understanding and scrutiny
- To describe how societies, industries and research use science

1. Choose the best description of a nebula. Tick **one** box.

☐ **A** A solar system that is forming

☐ **B** A very large star just before it explodes

☐ **C** The object found at the centre of a galaxy

☐ **D** A cloud of interstellar dust and gas

2. Name the force that causes nebulae to form stars and planets. Tick **one** box.

☐ **A** Gravity

☐ **B** Magnetism

☐ **C** Friction

☐ **D** Upthrust

3. Sort the following statements into the correct order to describe how a nebula can be formed from a star. Write the numbers **1** to **5** in the boxes. The first one has been done for you.

☐ **A** The star explodes as a supernova.

☐ 1 **B** A very large star runs out of the gas it uses to produce light.

☐ **C** Layers of gas and dust are thrown off to form a nebula.

☐ **D** The star expands to form a red supergiant.

☐ **E** The star collapses suddenly.

4. Name the **two** elements that make up most of the matter in a nebula. Tick **two** boxes.

☐ **A** Carbon

☐ **B** Helium

☐ **C** Silicon

☐ **D** Oxygen

☐ **E** Hydrogen

5. Name the **three** elements that make up most of the matter found in interstellar dust. Tick **three** boxes.

☐ **A** Carbon

☐ **B** Helium

☐ **C** Silicon

☐ **D** Oxygen

☐ **E** Hydrogen

Questions **6** to **8** refer to the following picture of a nebula.

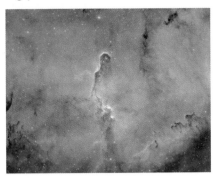

6. **a** Describe what a nebula contains.

_____

_____

Practical **b** A nebula does not produce light. Explain how we are able to observe a nebula.

_____

_____

7. Astronomers have observed a number of stars in this nebula that are around 100 000 years old.

a Choose the **best** description of the stage these stars have reached in their lives. Tick **one** box.

☐ **A** Recently formed

☐ **B** About the middle of their lives, similar to our Sun

☐ **C** Nearing the end of their lives

☐ **D** Recently exploded as supernovae

b Describe how these stars may have formed from the nebula.

_____

_____

_____

8. When the nebula shown in the picture was first observed, astronomers could only detect one very bright star close to the nebula, but no other stars within it. In 2003, astronomers were able to see for the first time a number of smaller, dimmer stars.

a Suggest why it took until 2003 for astronomers to observe the smaller, dimmer stars.

_____

_____

b Describe how the new evidence found in 2003 changed the model of what was happening in the nebula.

_____

_____

Questions **9** and **10** refer to the following picture that shows the closest nebula to Earth, called the Helix Nebula. It is a roughly circular cloud of gas and dust with a dim star at the centre.

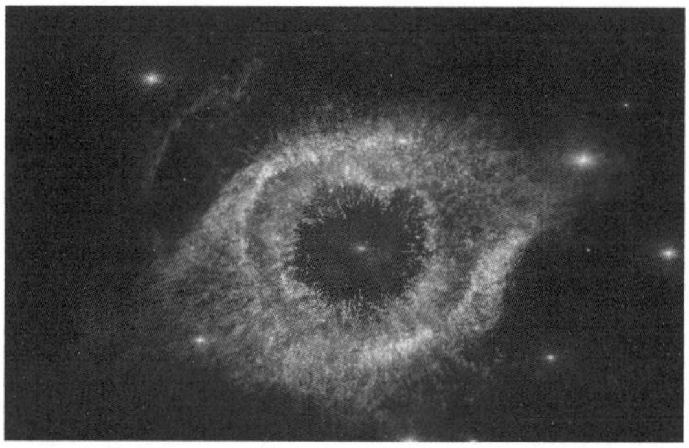

9.   **a** State what type of nebula this is: produced by an old star, or where new stars are forming.

_____

b Suggest **two** pieces of evidence to support your answer to part **a**.

1. _____

2. _____

c Predict how this nebula will change over long periods of time. Suggest what evidence could be collected to support your prediction.

_____

_____

_____

10. The Helix Nebula has been estimated to be 655 light-years from Earth.

Challenge **a** What does this measurement tell us about the light produced by the central star in the nebula?

_____

_____

b Why can this measurement only be an estimate?

_____

_____

c Suggest whether this nebula is inside or outside our galaxy, the Milky Way. Explain your answer.

_____

_____

Self-assessment

Tick the column which best describes what you know and what you are able to do.

What you should know:	I don't understand this yet	I need more practice	I understand this
The collision theory suggests that an asteroid colliding with the Earth caused the formation of the Moon			
The impact of an asteroid on Earth can cause many effects, including an explosion, shock waves, tsunami, climate change, extremely high winds, ground shaking and clouds of dust and gas			
Around 65 million years ago, about 75% of the species on Earth became extinct, including the dinosaurs			
The best hypothesis to explain this extinction is that an asteroid collided with Earth			
Nebulae are clouds of interstellar dust and gas			
Nebulae mostly contain the gases hydrogen and helium			
Nebulae can be formed by exploding massive stars at the end of their lives			
Dust and gas in some nebulae are slowly being pulled together by the force of gravity, eventually forming new stars powered by nuclear fusion of hydrogen			

You should be able to:	I can't do this yet	I need more practice	I can do this by myself
Describe the evidence for the collision theory of the formation of the Moon			
Describe the evidence for the impact of an asteroid causing mass extinctions			
Use models to understand past asteroid collisions and predict future collisions			
Organise the processes of a star's life into a cycle			
Describe how to observe nebulae to determine whether they are star-forming areas or have been produced by exploding stars			
Describe how evidence supports or contradicts a hypothesis or prediction, and how to amend a hypothesis when new evidence appears			
Understand that models and analogies can change according to scientific evidence			
Use observations to classify phenomena			
Interpret results and form conclusions using scientific knowledge			
Use scientific understanding to evaluate issues			
Discuss the development of scientific knowledge through scrutiny over time			
Describe some uses of science in research			

If you have ticked 'I don't understand this yet' or 'I can't do this yet' or mostly 'I need more practice', have another look at the relevant pages in the Student's Book. Then make sure you have completed all the questions in this Workbook chapter and the review questions in the Student's Book. If you have already completed all the questions ask your teacher for help and suggestions on how to progress.

Teacher's comments

End-of-chapter questions

....

1. Choose the generally accepted explanation for formation of the Moon. Tick **one** box.

☐ **A** A larger planet broke apart to form the Earth and the Moon.

☐ **B** A large asteroid collided with the Earth, causing parts of the Earth and the asteroid to be pulled together by gravitational forces to form the Moon.

☐ **C** A large asteroid was gradually pulled towards the Earth by gravitational forces as the Solar System formed, until it settled in orbit around the Earth.

☐ **D** Many smaller asteroids were pulled together by gravitational forces, then pulled into orbit around the Earth.

2. This question is about evidence for the most widely accepted explanation for the formation of the Moon.

a Describe **three** features of the Earth and of the Moon that are different.

_____

_____

_____

b Some rock samples taken from the Moon contain rocks that are very similar to rocks found on Earth. Explain what this evidence suggests about the formation of the Moon.

_____

_____

c List **two** types of evidence that scientists have found to support this theory.

_____

_____

_____

3. Some nebulae are described as 'stellar nurseries'.

a Describe how stars form from nebulae.

_____

_____

_____

b Name the process that takes place within stars, in which atoms join together and produce large amounts of energy and light.

_____

c Name the **two** elements that are used as fuel in this process.

_____   _____

4. The Alvarez hypothesis suggests an event occurred on Earth in the past, and that this is the reason why there are no dinosaurs on Earth today.

a Describe the Alvarez hypothesis.

_____

_____

b State how long ago this is thought to have occurred.

_____

c The effect of this event was to cause a mass extinction of dinosaurs and many other species of animals. Explain what 'mass extinction' means.

_____

d Explain how this event would have affected the Earth's climate.

_____

_____

e There is also evidence that plants were affected by this event. Describe this evidence.

_____

_____

5. 'Most of the hydrogen and helium gas found in space is recycled.'
Explain why this statement is correct.

Challenge

_____

_____

_____

_____

_____

_____